Springer Theses

Recognizing Outstanding Ph.D. Research

Aims and Scope

The series "Springer Theses" brings together a selection of the very best Ph.D. theses from around the world and across the physical sciences. Each volume has been selected for its scientific excellence by an internationally renowned institute or department. For greater accessibility to non-specialists, the published versions include an extended introduction, as well as a foreword by the student's supervisor explaining the special relevance of the work for the field. The series as a whole provides a valuable resource both for newcomers to the research fields, and for other scientists seeking detailed background information on special questions. Finally, it serves as an accredited documentation of the valuable contributions made by today's younger generation of scientists.

How to Apply

Theses prepared at internationally leading universities and institutes may be nominated for publication by supervisors with the endorsement of the head of department. Each such department is entitled to nominate one thesis per year. Enquiries and requests for a nomination form should be sent, together with the thesis, to Ms. Nathalie Berg (see link on right).

Eligibility Criteria

To qualify for consideration, theses should fulfil the following criteria:

- They must be written in good English.
- The topic should fall within the confines of Chemistry, Physics, Earth Sciences, Engineering and related interdisciplinary fields such as Materials, Nanoscience, Chemical Engineering, Complex Systems and Biophysics.
- The work reported in the thesis must represent a significant scientific advance.
- A maximum of 30% of the text may be a verbatim reproduction from the author's previous publications.
- They must have been examined and passed during the 12 months prior to nomination.
- The theses should have a clearly defined structure including an introduction accessible to new PhD students and scientists not expert in the relevant field.
- After acceptance, theses should be supplemented by a "Supervisor's Foreword" outlining the wider scientific relevance of the work

Indexed by zbMATH.

Ciaran Mark Hasnip

DUNE-PRISM—A New Method to Measure Neutrino Oscillations

Doctoral Thesis accepted by
Johannes Gutenberg-Universität, Mainz, Germany

Springer

Author
Dr. Ciaran Mark Hasnip
European Organization for Nuclear Research
Geneve, Switzerland

Supervisor
Prof. Alfons Weber
Johannes Gutenberg-Universität
Mainz, Germany

ISSN 2190-5053 ISSN 2190-5061 (electronic)
Springer Theses
ISBN 978-3-032-07250-4 ISBN 978-3-032-07251-1 (eBook)
https://doi.org/10.1007/978-3-032-07251-1

This Springer imprint is published by the registered company Springer Nature Switzerland AG
The registered company address is: Gewerbestrasse 11, 6330 Cham, Switzerland

Supervisor's Foreword

Neutrinos are peculiar particles. They have no charge, almost no mass, and barely interact with anything. Although they are the most abundant matter particles in the universe, many of their properties remain unknown.

The existence of the neutrino was first proposed in 1930 by Wolfgang Pauli to explain apparent violations of energy and angular momentum conservation in nuclear beta decay. Pauli suggested that an undetectable, massless particle could account for these discrepancies. However, neutrinos turned out not to be undetectable after all. Twenty-six years later, in a groundbreaking experiment at the Savannah River nuclear reactor, Clyde Cowan and Frederick Reines provided the first experimental confirmation of neutrinos, proving that they can indeed be observed.

Over the years, research on these elusive particles has been recognized with multiple Nobel Prizes. Scientists have discovered that neutrinos come in three different types and that, contrary to Pauli's original assumption, they do have mass—albeit incredibly tiny, at least a million times lighter than electrons. Neutrinos play a crucial role in understanding the evolution of the universe, as well as the creation and transformation of particles and chemical elements through the weak interactions.

Over the past few decades, neutrino experiments have been conducted using a variety of sources, including nuclear reactors, the Sun, cosmic rays interacting with the atmosphere, particle accelerators, and most recently, even particle colliders. While most experimental results align with the predictions of the Standard Model of particle physics, some anomalies remain unexplained.

One such anomaly, now well understood, is neutrino oscillation—the ability of neutrinos to change from one type to another. This phenomenon arises because the mass and weak interaction eigenstates of neutrinos are not identical. Over the last three decades, experiments have made increasingly precise measurements of the parameters governing these oscillations, including the differences in the neutrino mass squared, a set of mixing angles, and a so-called CP-violating phase. This CP-violating phase could cause neutrinos and antineutrinos to oscillate differently, potentially shedding light on why the universe contains more matter than antimatter. However, no definitive measurement of this phase exists yet, and a new generation of experiments is being developed to measure it with high precision.

One such experiment is the Deep Underground Neutrino Experiment (DUNE), which utilizes the world's most intense muon (anti-)neutrino beam, produced at Fermilab in Ilisonois, USA. DUNE consists of two detector systems: near detectors, located close to the accelerator to measure the neutrino beam composition before oscillations occur, and far detectors, located 1300 km away in South Dakota at the Sanford Underground Research Facility, buried one mile underground. DUNE is expected to provide the definitive measurement of CP violation in the neutrino sector.

To achieve such precise measurements, it is essential not only to understand the generated neutrino flux but also to accurately determine neutrino interaction cross-sections with detector materials. The latter remains a significant source of uncertainty, particularly in relating the true neutrino energy to the signals detected. These uncertainties arise due to complex nuclear medium effects.

As part of the DUNE collaboration, Ciaran Hasnip developed an innovative method to address these limitations. Using a movable near-detector system, he devised techniques to minimize reliance on neutrino interaction cross-section models. By positioning the near detector at different locations off the neutrino beam axis, he measured events under different neutrino fluxes and used this data to directly predict the far detector spectrum under different oscillation hypotheses. This approach has the potential to significantly reduces systematic uncertainties.

The theses begins with an introduction to neutrino physics and the motivation behind this research, followed by a detailed description of DUNE. It then presents the DUNE-PRISM concept, which enables direct prediction of the far detector oscillated data based on near detector measurements. A comprehensive analysis of the systematic uncertainties of this method follows, evaluating its effectiveness in determining oscillation parameters, including the critical CP-violating phase. The theses also explores case studies where traditional methods would produce significantly biased results. Finally, it concludes with an outlook on potential improvements.

This theses serves as a blueprint for future accelerator-based neutrino oscillation experiments. It introduces the subject clearly, dives into the technical complexities of the method, and provides a valuable reference for both beginners and experts in the field.

Mainz, Germany Prof. Alfons Weber
March 2025

Abstract

The Deep Underground Neutrino Experiment (DUNE) is a next-generation long baseline neutrino oscillation experiment designed to make precision measurements in a 1.2–2.4 MW neutrino beam, which is directed 1285 km from the Fermi National Accelerator Laboratory (Fermilab) to the Sandford Underground Research Facility (SURF) in South Dakota. Neutrinos are measured at two detector facilities: a near detector located at Fermilab close to where the beam is produced and a far detector at SURF. The neutrino beam can be configured to be composed primarily of either ν_μ or $\bar{\nu}_\mu$. DUNE measures the disappearance of ν_μ and $\bar{\nu}_\mu$ and the appearance ν_e and $\bar{\nu}_e$ in the neutrino beam. Measuring these neutrino flavour transitions provides DUNE with sensitivity to the neutrino mass ordering, δ_{CP}, θ_{13}, θ_{23} and the magnitude of Δm^2_{32}.

The DUNE Precision Reaction Independent Spectrum Measurement (DUNEPRISM) concept presents a novel way to perform a neutrino oscillation analysis, which has the potential to significantly reduce the impact of large systematic uncertainties in the neutrino interaction model. The PRISM method linearly combines measurements of off-axis neutrino interactions at the DUNE near detector to produce data-driven predictions of the oscillated neutrino event rate spectrum at the far detector. By building an oscillated far detector prediction directly from data, any unknown or poorly modelled neutrino interaction effects will be naturally incorporated into the measurement of the parameters of the neutrino oscillation model.

This thesis presents the first complete neutrino oscillation analysis for DUNE using the PRISM method. Details of the methodology are fully explained and the prospects for further improvements to the techniques described are highlighted. The expected impact and relative importance of the neutrino flux, cross section and detector systematic uncertainties are described in detail. Finally, this thesis demonstrates that the PRISM method is capable of performing a measurement of the oscillation parameters that is robust against neutrino interaction modelling errors.

Acknowledgements

Despite the best efforts of a global pandemic beginning in my first year, my D.Phil. has been an immensely rewarding and enjoyable experience. I feel very privileged to have had the opportunity to meet and work alongside researchers who are both fantastic physicists and wonderful people. Writing this thesis has only been possible thanks to their help and the support of many others.

First and foremost I would like to thank Alfons Weber for being a brilliant supervisor throughout my D.Phil. Alfons has taught me how to interpret my results, communicate my research clearly and provided me with a greater appreciation and understanding of physics. This thesis has been possible thanks to the constant support, advice and opportunities provided by Alfons.

I would also like to especially thank Giles Barr for becoming my second supervisor during the COVID-19 pandemic and providing such helpful practical and intellectual advice and support. Thank you as well to the other members of the Oxford accelerator neutrino group—Dave, Kirsty, Phil, Babak and Xianguo—who have always been generous in offering their help and guidance. A large part of what makes research enjoyable is working alongside your peers. I would like to thank my fellow students in the Oxford group: Soniya, Tom, Dan (now a postdoc at Oxford), John, Federico, Weijun, Kang, Lars, Aidan, Fabio and Laurence for their friendship and making the office a fun place to work over the last four years—even if it was a Zoom office for a long time! Soniya, your company and our after-work Wendy's dinners helped me through some of the most difficult parts of my D.Phil. In return, I promise that I will one day learn a salsa step. As a mark of our friendship, I now give full credit to Tom for providing the creative, if slightly uninformative, unofficial thesis title: "PRISM: the Dark Side of the DUNE".

I would like to acknowledge the work of everyone in the DUNE-PRISM working group that has been so important in enabling this thesis. I would like to especially thank the convener of the DUNE-PRISM working group, Mike Wilking, for guiding my research and providing me with many fantastic opportunities, such as my long term attachment at Stony Brook. Thank you to everyone in the Stony Brook Neutrino and Nucleon decay group for welcoming me as a visiting researcher; I hope you can achieve greater success in future physics football tournaments than you did with me on the pitch. I greatly enjoyed working with Wei Shi, who made a lot of effort to make me feel welcome upon moving to Long Island.

Thank you to the conveners of the Long Baseline working group, Luke Pickering and Callum Wilkinson, who both did so much to teach me how to do a neutrino oscillation analysis. Luke always encouraged me in my research and was there to help with my many coding issues, as well as provide a ready supply of memes. Callum also took the time to produce DUNE sensitivities that are used in this thesis for contour comparisons.

I would like to thank all of my friends and family, especially my parents, for supporting me and keeping me sane during a challenging four years. Finally, a special thank you to Eloise for being so supportive of my efforts, particularly throughout the pandemic and my six months away on Long Island; I am eternally grateful for your patience, generosity and kindness.

Contents

1 Introduction 1
References 4

2 Neutrino Theory and Background 5
2.1 Neutrinos: A Brief History 5
2.2 Neutrinos in the Standard Model 7
2.2.1 The Weak Interaction 8
2.3 Neutrino Oscillations 9
2.3.1 Neutrino Oscillations of Two Flavours 10
2.3.2 Neutrino Oscillations of Three Flavours 11
2.3.3 Neutrino Oscillations in Matter 13
2.4 Neutrino Oscillation Measurements 14
2.4.1 Solar Neutrinos 14
2.4.2 Atmospheric Neutrinos 15
2.4.3 Reactor Neutrinos 17
2.4.4 Accelerator Neutrinos 18
2.4.5 Overview of Oscillation Parameter Measurements 20
2.5 Neutrinos as a Window to New Physics 22
2.6 Modelling Neutrino Interactions 23
References 27

3 Deep Underground Neutrino Experiment 33
3.1 Liquid Argon Time Projection Chamber 33
3.1.1 Particle Interactions in Liquid Argon 35
3.1.2 Detector Effects in Liquid Argon 37
3.1.3 Scintillation Light in Liquid Argon 38
3.2 Producing a Neutrino Beam for DUNE 39
3.3 DUNE Far Detector 41
3.3.1 Horizontal Drift Module 41
3.3.2 Vertical Drift Module 43

3.3.3 Data Acquisition and Triggering ... 44
3.3.4 Calibration ... 44
3.4 DUNE Near Detector Complex ... 45
3.4.1 Near Detector Liquid Argon ... 46
3.4.2 Downstream Tracker ... 47
3.4.3 SAND ... 48
3.5 PRISM ... 49
3.6 Monte Carlo Data ... 50
3.6.1 Far Detector Simulation ... 51
3.6.2 Far Detector Reconstruction ... 53
3.6.3 Near Detector Simulation ... 54
3.6.4 Near Detector Parameterised Reconstruction ... 55
References ... 56

4 Data-Driven Far Detector Prediction ... 59
4.1 Data Selection ... 59
4.1.1 Analysis Variables ... 59
4.1.2 Far Detector CVN Selection ... 61
4.1.3 Near Detector Selection ... 64
4.1.4 Additional Horn Current Sample ... 65
4.2 Near Detector Run-Plan ... 67
4.3 Correcting for Detector Effects ... 68
4.3.1 Near Detector Background Subtraction ... 69
4.3.2 Removing Near Detector Effects ... 70
4.3.3 $\sigma(\nu_e)/\sigma(\nu_\mu)$ Cross Section Ratio ... 73
4.3.4 Introduce Far Detector Effects ... 74
4.4 Linear Combination ... 77
4.5 Far Detector Monte Carlo Corrections ... 78
4.6 PRISM Predictions ... 80
4.6.1 Predictions for any Oscillation Parameter Hypothesis ... 83
4.7 Future Methodology Improvements ... 86
4.7.1 Geometric Efficiency Correction ... 86
4.7.2 Linear Combination Predictions of the Far Detector Backgrounds ... 89
4.7.3 PRISM $\sigma(\nu_e)/\sigma(\nu_\mu)$ Measurement ... 90
4.7.4 Alternative Analysis Variables ... 90
References ... 92

5 Systematic Uncertainties for PRISM ... 93
5.1 Flux Uncertainties ... 93
5.1.1 Flux Uncertainty Cancellation ... 96
5.1.2 Flux Uncertainties in a PRISM Measurement ... 100
5.2 Cross Section Uncertainties ... 104
5.2.1 Cross Section Parameter Impact ... 107
5.2.2 Near Detector Efficiency Calculation ... 108

5.3 Detector Uncertainties 110
References 117

6 Four-Channel PRISM Oscillation Analysis 119
6.1 Exposure at the Near and Far Detectors 119
6.2 NuFIT 4.0 Oscillation Parameters 120
6.3 MINUIT Oscillation Fit 121
6.4 PRISM Goodness of Fit 122
6.5 PRISM Sensitivities 124
6.5.1 Chi-Square Parameter Scans 125
6.5.2 Statistics-Only Sensitivities 125
6.5.3 Sensitivities with Systematic Uncertainties 127
6.6 Impact of Systematic Uncertainties 128
6.6.1 Flux Systematic Uncertainty Impact 130
6.6.2 Cross Section Uncertainties and the Near Detector Efficiency 137
6.6.3 Detector Systematic Uncertainty Impact 138
6.7 PRISM Sensitivities with a Data-Driven Near Detector Efficiency Calculation 143
6.8 Comparison with On-Axis Sensitivities 145
6.9 Summary of PRISM Sensitivities 153
References 154

7 Oscillation Measurement Biases from the Neutrino Interaction Model 155
7.1 Deposited Energy Analysis Variable 156
7.2 Missing Proton Energy Mock Data 156
7.2.1 Mock Data Impact on Oscillation Measurement 158
7.3 PRISM Oscillation Measurement with Mock Data 158
7.3.1 Data-Driven Geometric Efficiency Correction 163
7.3.2 Wrong-Sign Background 164
7.4 Mock Data Fits Accounting for Systematic Uncertainty 167
References 169

8 Conclusions and Outlook 171
References 173

Appendix: Tikhonov Regularisation Optimisation 175

Abbreviations

APA	Anode Plane Assembly
CC	Charged Current
CCQE	Charged Current Quasielastic
CPA	Cathode Plane Assembly
CVN	Convolutional Visual Network
DAQ	Data Acquisition
DIS	Deep Inelastic Scattering
DUNE	Deep Underground Neutrino Experiment
FD	Far Detector
FHC	Forward Horn Current
GENIE	Generates Events for Neutrino Interaction Experiments
LBNF	Long Baseline Neutrino Facility
LArTPC	Liquid Argon Time Projection Chamber
MC	Monte Carlo
NC	Neutral Current
ND	Near Detector
ND-LAr	Near Detector Liquid Argon Detector
ND-GAr	Near Detector Gas Argon Detector
POT	Protons on Target
PPFX	Package to Predict the FluX
PRISM	Precision Reaction Independent Spectrum Measurement
RHC	Reverse Horn Current
SAND	System for on-Axis Neutrino Detection
SiPM	Silicon Photomultiplier
SNB	Supernova Neutrino Burst
TMS	Temporary Muon Spectrometer
VUV	Vacuum Ultraviolet

List of Figures

Fig. 2.1 Charged current (CC) and neutral current (NC) scattering of a neutrino off a nucleon. Here, $N = p, n$ 9
Fig. 2.2 The resonance of the Z^0 as measured by LEP. The width is consistent with exactly three flavours of weakly interacting neutrinos. Reprinted from Phys. Rep., vol. 427 (5–6), The ALEPH Collaboration et al., "Precision Electroweak Measurements on the Z Resonance", pp. 257–454, Copyright (2006), with permission from Elsevier [44] 10
Fig. 2.3 $l = e, \mu$ or τ and $N = e, p$ or n 13
Fig. 2.4 Ratio of the measured background and geoneutrino subtracted $\bar{\nu}_e$ event rate to the expected $\bar{\nu}_e$ in the case of no oscillations as a function of L_0/E in the KamLAND experiment. L_0 is the effective baseline calculated from the flux-weighted average distance of all the reactors. Reprinted figure with permission from S. Abe et al. (KamLAND Collaboration), Phys. Rev. Lett., vol. 100, p. 221803, 2008. Copyright (2008) by the American Physical Society [55] 16
Fig. 2.5 Atmospheric neutrino results from Super-Kamiokande. The sub-GeV and multi-GeV μ-like and e-like events are plotted as a function of the zenith angle. The data points are plotted alongside the Monte Carlo prediction assuming no neutrino oscillations in the hatched boxes. A line is fitted to the data that assumes $\nu_\mu \to \nu_\tau$ oscillations occur. Reprinted figure with permission from Y. Fukuda, et al. (Super-Kamiokande Collaboration), Phys. Rev. Lett., vol. 81, pp. 1562–1567, 1998. Copyright (1998) by the American Physical Society [31] 17

Fig. 2.6 Fully reconstructed muon neutrino event rate at the MINOS far detector and the expected event rate in the case of no oscillations. The bottom plot is the ratio of the background subtracted data to the no oscillations hypothesis prediction. Reprinted figure with permission from P. Adamson, et al. (MINOS Collaboration), Phys. Rev. Lett., vol. 106, p. 181801, 2011. Copyright (2011) by the American Physical Society [66] 19
Fig. 2.7 Diagram illustrating the two scenarios for the ordering of the neutrino masses. Either ν_3 is the heaviest state (normal ordering) or the lightest state (inverted ordering). ©IOP Publishing. Reproduced with permission. All rights reserved [74] 20
Fig. 2.8 Oscillation probabilities for neutrinos and antineutrinos in the Earth's crust, assuming normal mass ordering and a 1300 km baseline. Probabilities for different values of δ_{CP} are shown. The oscillation probability for neutrinos at the first oscillation maximum (at approximately 2.5 GeV) is enhanced relative to the antineutrino case. Figure taken from Reference [75] 21
Fig. 2.9 Current status of measurements of Δm^2_{32} and $\sin^2\theta_{23}$. The precision of the measurement of these parameters is currently led by the long baseline experiments. Reprinted from Progress in Particle and Nuclear Physics, vol. 124, M. S. Athar et al., "Status and perspectives of neutrino physics," pp. 103947, Copyright (2022), with permission from Elsevier [78] 22
Fig. 2.10 Feynman diagrams for the three main neutrino interaction types relevant to long baseline neutrino oscillation experiments 24
Fig. 2.11 Total muon neutrino CC cross section per nucleon. Contributions from different interaction types are shown. CCQE events dominate at low energies and DIS dominates at high energies. Reprinted figure with permission from J. A. Formaggio and G. P. Zeller, Rev. Mod. Phys., vol. 84, pp. 1307–1341, 2012. Copyright (2012) by the American Physical Society [93] 25

Fig. 2.12 Illustration of how FSI effects can alter the kinematics and multiplicities of observable hadrons in the detector. Pions are often absorbed by the nuclear medium, potentially biasing the neutrino energy estimation. Figure produced by Tomasz Golan. Reprinted from Progress in Particle and Nuclear Physics, vol. 100,L. Alvarez-Ruso et al. (Neutrino Scattering Theory Experiment Collaboration), "NuSTEC White Paper: Status and challenges of neutrino–nucleus scattering," pp. 1–68, Copyright (2018), with permission from Elsevier [88] 26
Fig. 3.1 Schematic diagram of a LArTPC [4]. The diagram shows a LArTPC with a horizontal electron drift direction and wire-plane electronic readout design. ©IOP Publishing. Reproduced with permission. All rights reserved 34
Fig. 3.2 LArTPC event displays of neutrino events in the MicroBooNE detector [9]. The muon track is the longest of the three tracks originating from the interaction vertex in Fig. 3.2a. The electromagnetic shower in Fig. 3.2b is produced at the interaction vertex, resulting in many discrete charge depositions across the length of the event display 35
Fig. 3.3 Mean energy loss in several different media as a function of muon, pion and proton momenta according to the Bethe-Bloch equation. Figure taken from Reference [10] 36
Fig. 3.4 Diagrams illustrating how an electromagnetic shower is produced. Electrons and positrons radiate photons, which subsequently pair produce 37
Fig. 3.5 Schematic diagram of the long baseline neutrino facility (LBNF). Figure from Ref. [20] 40
Fig. 3.6 Illustration the LBNF target hall and decay pipe. Protons enter from the left through a beryllium "window" before passing through the baffle. Protons interact with the target that is surrounded by Horn A. Hadrons are subsequently focused by Horn B and C before entering the decay pipe. The proton beam is tilted down by 0.101 radians. A hadron absorber stops any remaining hadrons in the beamline 40
Fig. 3.7 Schematic diagram of the FD detector module from Ref. [25] 41
Fig. 3.8 Schematic diagram of the FD horizontal drift TPC from Ref. [25] 42

Fig. 3.9 Schematic diagram of the horizontal drift model APA frame from Ref. [25]. The different wire planes are drawn in addition to the cold electronics boards highlighted in blue on the right edge of the APA 43
Fig. 3.10 Schematic diagram of the ND detector complex. The diagram shows ND-LAr and ND-GAr in the far off-axis position at 28.5 m from the beam axis centre [33] 45
Fig. 3.11 Schematic diagram of the ND-LAr module layout 47
Fig. 3.12 Schematic diagram of SAND. Figure created and copyright owned by Matteo Vicenzi [41] 49
Fig. 3.13 Pions are produced when protons hit the LBNF target. On the left is the neutrino energy as a function of the pion energy for different angles relative to the neutrino beam axis. The right plot shows the corresponding neutrino flux spectra for the same four angles. At high angles the neutrino energy becomes relatively constant with pion energy, resulting in an approximately monochromatic neutrino flux 49
Fig. 3.14 Fluxes simulated in the FD and ND by G4LBNF for different neutrino flavours. FD fluxes in neutrino enhanced and anti-neutrino enhance mode for ν_μ, $\bar{\nu}_\mu$, ν_e and $\bar{\nu}_e$ fluxes shown in Fig. 3.14a, b. G4LBNF generates neutrino fluxes at the ND both on-axis and going up to 40 m off-axis. Shown are the ν_μ and $\bar{\nu}_\mu$ fluxes generated at all off-axis positions in FHC and RHC mode respectively 51
Fig. 3.15 Illustration of the three types of FD flux files that allow for three-flavour oscillations for any set of parameters. The "non-swap" files do not change any neutrino flavour, whilst the "electron-swap" and "tau-swap" do change the initial neutrino flavours 52
Fig. 3.16 Three reconstructed hits fitted with a Gaussian from the ProtoDUNE Single-Phase. Figure by the DUNE Collaboration [49]. Published by IOP Publishing Ltd on behalf of Sissa Medialab 53
Fig. 3.17 Side view of the simulated ND complex. On the left is ND-LAr and on the right is ND-GAr. The neutrino beam axis is shown in red passing through the centre of both detectors [1] 56
Fig. 4.1 Flow diagram summarising the process of producing a PRISM prediction of the ν_μ, $\bar{\nu}_\mu$, ν_e and $\bar{\nu}_e$ FD event rate. The start of the process is the top left box. The blue box is only performed for ν_e and $\bar{\nu}_e$ appearance predictions 60

Fig. 4.2 CVN scores for FD selection of ν_μ ($\bar{\nu}_\mu$) and ν_e ($\bar{\nu}_e$) in FHC (RHC) mode. Figures from Ref. [2]. The red line shows the position of the cut applied on the CVN score of each event ... 62

Fig. 4.3 Event rate spectra for selected ν_μ, $\bar{\nu}_\mu$, ν_e and $\bar{\nu}_e$ events in the FD. The stacked histograms show the individual signal and background components of the total selected event rate for each signal channel. The event rates assume the NuFIT 4.0 oscillation parameters [3] ... 63

Fig. 4.4 Event rate spectra for selected ν_μ and $\bar{\nu}_\mu$ events in the ND. The ND event rate spectra in Fig. 4.4a and b are a function of visible $E^{rec.}_{ND}$ and off-axis position ... 65

Fig. 4.5 Neutral current and wrong-sign ND background event rate in the in FHC and RHC mode ... 66

Fig. 4.6 Ratios of the 280 kA to the 293 kA LBNF flux at the ND for ν_μ and $\bar{\nu}_\mu$ in FHC and RHC mode respectively ... 66

Fig. 4.7 ND event rates on-axis for the 280 kA horn current sample ... 67

Fig. 4.8 ND run-plan and ND event rates in FHC and RHC mode weighted for the chosen run-plan. The ND run plan is given in the number of weeks at each position in the assumed 28-week beam year (**a**) and the corresponding weights for each off-axis position (**b**). The small on-axis ND sample taken at the lower 280 kA horn current is also included in the run plan ... 68

Fig. 4.9 Background subtracted ND signal for the chosen run-plan and both horn polarities. Run-plan weighted predicted background events that were subtracted from the ND data also shown ... 69

Fig. 4.10 ND selection efficiency at each off-axis position as a function of $E^{true}_{vis.}$ and ND smearing matrices normalised to the efficiency at the on-axis position ... 71

Fig. 4.11 Cross section ratios for the electron neutrino appearance analysis in FHC and RHC mode as a function of $E^{vis.}_{true}$. The error bars originate from the statistical uncertainty on the MC predictions used to calculate the ratios ... 73

Fig. 4.12 FD ν_μ, $\bar{\nu}_\mu$, ν_e and $\bar{\nu}_e$ selection efficiency calculated as a function of $E^{vis.}_{true}$ ... 74

Fig. 4.13 FD ν_μ, $\bar{\nu}_\mu$, ν_e and $\bar{\nu}_e$ smearing matrices normalised to the selection efficiencies shown in Fig. 4.12 ... 75

Fig. 4.14 ND data that has corrections applied to account for differences between the ND and FD ... 76

Fig. 4.15 Linear combination coefficients calculated by matching ND off-axis fluxes to the FD flux. Shown are the coefficients for the 293 kA and 280 kA ND flux samples ... 79

Fig. 4.16 FD flux prediction and the linearly combined ND fluxes. The residual is the fractional difference between the FD flux and the linearly combined ND fluxes 80
Fig. 4.17 Corrections from the FD MC that include backgrounds and the flux miss-match correction. The FD exposure is 50 kt-MW-Yrs 81
Fig. 4.18 PRISM predictions corresponding to set of linear combination coefficients shown in Fig. 4.15 for the four DUNE signal channels mode. The total stacked histogram is the PRISM prediction. The FD exposure is 50 kt-MW-Yrs ... 82
Fig. 4.19 Covariance matrices for ν_μ, $\bar{\nu}_\mu$, ν_e and $\bar{\nu}_e$ predictions. The covariance matrices are associated with the PRISM predictions shown in Fig. 4.18. The diagonal elements are the variance of each bin of the PRISM prediction 83
Fig. 4.20 PRISM disappearance predictions for three different oscillation hypotheses in FHC (top) and RHC (bottom) mode. Muon neutrino disappearance is dominated by Δm^2_{32} and $\sin^2\theta_{23}$ and the different values of these parameters for the three predictions are shown in the oscillation parameter space at the top right of each plot 84
Fig. 4.21 PRISM appearance predictions for three different oscillation hypotheses in FHC (top) and RHC (bottom) mode. Measurements of electron neutrino appearance are sensitive to δ_{CP} and $\sin^2 2\theta_{13}$ and the different values of these parameters for the three predictions are shown in the oscillation parameter space at the top right of each plot 85
Fig. 4.22 Sketch illustrating the geometric efficiency calculation. The neutrino event is rotated about the neutrino axis and translated in the y–z plane a number of times. The red box represents the fixed x position of the y–z plane in which the event is translated. Whether the hadronic veto cut is passed or failed is indicated. The muon selection probability is also shown. For the five random throws shown the geometric efficiency is therefore: $(0 \times 0.8 + 1 \times 0.3 + 0 \times 0.7 + 1 \times 0.4 + 1 \times 0.95)/5 = 0.33$ 88
Fig. 4.23 Total PRISM predictions of the FD event rate using the two-dimensional analysis variable $E_{had.}$ versus $E_{lep.}$ 91
Fig. 5.1 Fractional change in the ND true ν_μ CC event rate due to a 1σ increase in the Horn Current parameter. In addition to the large shift on-axis at 4 GeV, there are smaller shifts in the flux at off-axis positions not present on-axis 96

Fig. 5.2 Fractional change due to the Horn Current parameter in the FD MC oscillated event rate prediction and PRISM linear combination as a function of true neutrino energy. The PRISM prediction is generated by linearly combining the shifted ND event rates with coefficients calculated from the nominal MC. Similar fractional shifts can be seen in the FD MC and PRISM predictions in the 4 GeV region 97
Fig. 5.3 The fractional shifts in the FD MC and PRISM linear combination predictions due to the Horn Current parameter seen in Fig. 5.2 are plotted together and the difference between them is taken. The difference between the shifts in the FD MC and PRISM predictions can be assumed to be the total systematic uncertainty in the linear combination. Good systematic cancellation is seen at 4 GeV 98
Fig. 5.4 Difference between the fractional shifts in the FD MC and PRISM linear combination due to hadron production uncertainty PCA parameters 1–4 for each oscillation signal channel. The symbols $P^{\nu_l}_{\text{diff}}$ and $F^{\nu_l}_{\text{diff}}$ refer to the fractional difference of the PRISM and FD MC prediction due to the systematic shift respectively. As expected, the first hadron production PCA component has the most significant effect. Reasonable cancellation occurs at higher energies 99
Fig. 5.5 Difference between the fractional shifts in the FD MC and PRISM linear combination due hadron production uncertainty PCA parameters 5–8 for each oscillation signal channel. The symbols $P^{\nu_l}_{\text{diff}}$ and $F^{\nu_l}_{\text{diff}}$ refer to the fractional difference of the PRISM and FD MC prediction due to the systematic shift respectively. The higher PCA components have smaller impact than the first 4 PCA components shown in Fig. 5.4 100
Fig. 5.6 Difference between the fractional shifts in the FD MC and PRISM linear combination due to the focusing uncertainty parameters for each oscillation signal channel. The symbols $P^{\nu_l}_{\text{diff}}$ and $F^{\nu_l}_{\text{diff}}$ refer to the fractional difference of the PRISM and FD MC prediction due to the systematic shift respectively. The uncertainty in the decay pipe radius has a substantial impact due to the very large a priori uncertainty assigned to this parameter 101

Fig. 5.7 Difference between the fractional shifts in the FD MC and PRISM linear combination due to the horn alignment uncertainty parameters for each oscillation signal channel. The symbols $P^{\nu_l}_{\text{diff}}$ and $F^{\nu_l}_{\text{diff}}$ refer to the fractional difference of the PRISM and FD MC prediction due to the systematic shift respectively. The parameter Horn 1 X-Shift has an important effect due to changes in the flux off-axis that do not cancel with changes in the FD flux 102

Fig. 5.8 Difference between the fractional shifts in the FD MC and PRISM linear combination due to the beam alignment uncertainty parameters for each oscillation signal channel. The symbols $P^{\nu_l}_{\text{diff}}$ and $F^{\nu_l}_{\text{diff}}$ refer to the fractional difference of the PRISM and FD MC prediction due to the systematic shift respectively. The parameter Beam X-Offset has an important effect 103

Fig. 5.9 Fractional change in the ND true ν_μ CC event rate due to a 1σ increase in the Horn 1 X-Shift and Beam X-Offset parameters. The shifts observed at off-axis position in the ND fluxes for these parameter will not necessarily cancel with any change in the FD flux, leading to a larger contribution to the systematic uncertainty 103

Fig. 5.10 Comparison of the 1σ error band when all the MC components of the total PRISM prediction are able to vary and when the linear combination coefficient calculation is fixed. The magnitude of the 1σ systematic uncertainty band is plotted as a function of reconstructed neutrino energy for each oscillation channel. The magnitude of the 1σ error band for the ν_μ prediction reduces substantially when the linear combination coefficient calculation is fixed 104

Fig. 5.11 The magnitude of the 1σ systematic uncertainty band is plotted as a function of reconstructed neutrino energy for each oscillation channel and each category of GENIE cross section systematic uncertainties set out in Table 5.2. Each error band is the combined effect of all the parameters associated with each category of cross section uncertainties set out in Table 5.2 108

Fig. 5.12 The magnitude of the 1σ systematic uncertainty band is plotted as a function of reconstructed neutrino energy for each oscillation channel and each type of non-GENIE cross section systematic uncertainty set out in Table 5.3. The legend refers to the effect of the different cross section uncertainty types in Table 5.3. For example, the "Non-Res. DIS" error band includes the effect of all the non-resonant CC and non-CC DIS parameters 109

Fig. 5.13 The magnitude of the 1σ systematic uncertainty band is plotted as a function of reconstructed neutrino energy for each oscillation channel and for the cross section ratio uncertainties contained in Table 5.3. The cyan line is the same as in Fig. 5.12. However, for the red error band the calculation of the cross section ratio correction for the PRISM appearance channel (see Sect. 4.3.3 for details) is fixed. This substantially reduces the impact of the cross section ratio parameters 110
Fig. 5.14 Comparison of the 1σ error band when all the MC components of the total PRISM prediction are able to vary and when the ND efficiency calculation is fixed. The magnitude of the 1σ systematic uncertainty band is plotted as a function of reconstructed neutrino energy for each oscillation channel. The magnitude of the 1σ error band in generally reduces substantially when the ND efficiency correction is fixed 111
Fig. 5.15 Functional form of the energy scale uncertainty parameters p_0, p_1 and p_2 for each particle interaction type and the total energy scale. The combined effect is the sum of the p_0, p_1 and p_2 parameters 113
Fig. 5.16 Comparison of the 1σ error bands when varying the PRISM prediction by the p_0, p_1 and p_2 parameters for total energy scale uncertainty in the ND (red) and FD (blue) 114
Fig. 5.17 Comparison of the 1σ error band when varying the PRISM prediction by the individual detector response uncertainties. For example, the $\mu^\pm$ error band quantifies the collective impact of the p_0, p_1 and p_2 parameters for the muon energy scale uncertainty that is correlated between the ND and FD 115
Fig. 5.18 Comparison of the 1σ error band when varying the PRISM prediction by the resolution uncertainties in the ND (red) and FD (blue). Each error band is made by collectively varying all the resolution uncertainties in the corresponding detector 116
Fig. 5.19 Comparison of the 1σ error band when varying the PRISM prediction by all the detector uncertainties simultaneously. The two solid lines correspond to error bands where all the MC components are allowed to be varied by the detector systematic parameters (blue) and where the ND and FD smearing matrices are prevented from varying in the fit (red). The two dotted lines are when either the FD or ND smearing matrix is fixed. The size of the error band dramatically reduces when the smearing matrices are fixed 117

Fig. 6.1 Illustration of the agreement between the PRISM prediction to the FD data in an oscillation parameter space. The FD data in each window is set to the NuFIT 4.0 oscillation parameters. A PRISM prediction is generated at each point in the Δm^2_{32} and $\sin^2\theta_{23}$ parameter space and fitted to the FD data. At the true oscillation point (top left window) the PRISM prediction will have the best fit to the FD data, resulting in the smallest χ^2 calculation 122
Fig. 6.2 A series of "χ^2-scans" (see text for details) for Δm^2_{32}, $\sin^2\theta_{23}$, $\sin^2 2\theta_{13}$ and δ_{CP}. The y-axis is $\Delta\chi^2(\boldsymbol{\theta}) = \chi^2(\boldsymbol{\theta}) - \chi^2_{min.}$. The contribution to the measurement of the parameters from each of the oscillation signal channels is shown. The combined χ^2, calculated by summing the χ^2 calculation from the four signal channels, is also shown 126
Fig. 6.3 Contours for Δm^2_{32} versus $\sin^2\theta_{23}$ (top) and $\sin^2 2\theta_{13}$ versus δ_{CP} (bottom). The sensitivities include the oscillation parameters in the fit and statistical uncertainties, however no systematic uncertainty parameters have yet been included. The penalty term on θ_{13} is removed for the $\sin^2 2\theta_{13}$ and δ_{CP} contours 127
Fig. 6.4 Statistics-only fits for Δm^2_{32}, $\sin^2\theta_{23}$, $\sin^2 2\theta_{13}$ and δ_{CP} at two different exposures. The impact of the statistical uncertainty is demonstrated by removing the linear combination covariance matrix from the χ^2 calculation. The 1σ and 3σ contours are shown at the bottom of each plot. The total width of the bar is the 3σ contour and the small bars represent the 1σ contours. The true point is indicated by the vertical grey line 128
Fig. 6.5 PRISM oscillation sensitivities for three exposures and assuming the NuFIT 4.0 true oscillation parameters. All of the flux, cross section and detector uncertainties are accounted for in the fit as nuisance parameters in addition to the oscillation parameters. Penalty terms are included for θ_{12}, Δm^2_{21}, ρ and $\sin^2 2\theta_{13}$. Note that the 5σ discovery sensitivity is shown for δ_{CP} 129
Fig. 6.6 PRISM oscillation sensitivities when accounting for no systematic uncertainty, only the flux uncertainty, the flux and cross section uncertainty and finally all the sources of systematic uncertainty. This demonstrates the relative impact of the different types of systematic parameter on Δm^2_{32} (top), $\sin^2\theta_{23}$ (middle) and δ_{CP} (bottom) ... 131

Fig. 6.7 PRISM oscillation sensitivities when accounting for no systematic uncertainty and all the flux uncertainty parameters. In the case of the blue contours, the calculation of linear combination coefficients is not allowed to vary in the fit. All the MC components of the PRISM prediction are allowed to vary for the red contours 132
Fig. 6.8 PRISM oscillation sensitivities when accounting for no systematic uncertainty, only the hadron production uncertainty, the combined focusing and beam and horn alignment uncertainty and finally the total flux uncertainty. This demonstrates the relative impact of the different categories of flux parameter on Δm^2_{32} (top), $\sin^2\theta_{23}$ (middle) and δ_{CP} (bottom) 133
Fig. 6.9 PRISM oscillation sensitivities when accounting for no systematic uncertainty, only the focusing uncertainty, the beam alignment uncertainty and finally the horn alignment uncertainty. This demonstrates the relative impact of the different types of flux parameter on Δm^2_{32} (top), $\sin^2\theta_{23}$ (middle) and δ_{CP} (bottom) 135
Fig. 6.10 "N-1" systematic study (see the text for details) for Δm^2_{32}. Four different flux parameters (see the legends) are removed to test the relative contribution of that parameter to the overall systematic impact of a flux uncertainty category. The 1σ and 3σ contours are shown at the bottom of each $\Delta\chi^2$ curve 136
Fig. 6.11 "N-1" systematic study (see the text for details) for $\sin^2\theta_{23}$. Four different flux parameters (see the legends) are removed to test the relative contribution of that parameter to the overall systematic impact of a flux uncertainty category. The 1σ and 3σ contours are shown at the bottom of each $\Delta\chi^2$ curve 137
Fig. 6.12 Impact of accounting for the cross section uncertainties on the PRISM sensitivity in two cases. The blue line is the nominal case where all the MC components are affected by systemic uncertainties. The red line is the case where the ND efficiency correction calculation is not affected by systematic uncertainties. Compare the red and blue contours to the black dotted line, where only the flux uncertainties are accounted for 138
Fig. 6.13 PRISM oscillation sensitivities for Δm^2_{32}, $\sin^2\theta_{23}$ and δ_{CP} when accounting for the total energy scale uncertainties in the ND and FD. Each fit shows the combined effect of the p_0, p_1 and p_2 energy scale parameters 140

Fig. 6.14 PRISM oscillation sensitivities for Δm^2_{32}, $\sin^2\theta_{23}$ and δ_{CP} when accounting for the individual particle energy scale uncertainties. This includes the energy scale uncertainties for muon, charged hadrons, EM showers and neutrons. Each fit shows the combined effect of the p_0, p_1 and p_2 energy scale parameters 141
Fig. 6.15 PRISM oscillation sensitivities for Δm^2_{32}, $\sin^2\theta_{23}$ and δ_{CP} when accounting for the resolution uncertainty in the ND and FD. The coloured lines show the combined effect of all the resolution uncertainties in the ND (blue) and FD (red) 142
Fig. 6.16 PRISM oscillation sensitivities for three exposures and assuming the NuFIT 4.0 true oscillation parameters. All sources of systematic uncertainty (flux, cross section and detector) are accounted for. The calculation of the ND efficiency correction is fixed in all the oscillation fits. Note that the 5σ discovery sensitivity is shown for δ_{CP} 144
Fig. 6.17 PRISM oscillation contours at a fixed exposure in two-dimensional oscillation parameter spaces. All sources of systematic uncertainty (flux, cross section and detector) are accounted for. The reactor neutrino θ_{13} constraint is applied for both sensitivities 145
Fig. 6.18 PRISM oscillation contours for $\sin^2 2\theta_{13}$ and δ_{CP} at a fixed exposure in two-dimensional oscillation parameter spaces. All sources of systematic uncertainty (flux, cross section and detector) are accounted for. The reactor neutrino θ_{13} constraint is removed for the second (bottom) oscillation sensitivity 146
Fig. 6.19 Contours for a δ_{CP} PRISM measurement assuming a value of $\delta_{CP} = -0.5$. All systematic uncertainties are accounted for and the calculation of the ND efficiency correction is fixed 146
Fig. 6.20 Comparison between PRISM and On-Axis analysis contours for a fixed exposure and accounting for all sources of systematic uncertainty. The ND efficiency correction has been fixed in the PRISM oscillation fits, simulating the inclusion of a data-driven ND efficiency correction 148
Fig. 6.21 Comparison between the PRISM (red) and On-Axis (blue) oscillation contours at a fixed exposure in two-dimensional oscillation parameter spaces. The reactor neutrino θ_{13} constraint is removed for the $\sin^2 2\theta_{13}$ versus δ_{CP} sensitivity. All sources of systematic uncertainty are accounted for in the sensitivities. The ND efficiency correction has been fixed in the PRISM oscillation fits, simulating the inclusion of a data-driven ND efficiency correction 149

Fig. 6.22 Contours for Δm^2_{32} when accounting for the cross section (top), flux (middle) and detector (bottom) uncertainties. The impact of these systematic uncertainties on the PRISM (left) and On-Axis (right) contours are to be compared 150
Fig. 6.23 Contours for $\sin^2\theta_{23}$ when accounting for the cross section (top), flux (middle) and detector (bottom) uncertainties. The impact of these systematic uncertainties on the PRISM (left) and On-Axis (right) contours are to be compared 151
Fig. 6.24 Contours for δ_{CP} when accounting for the cross section (top), flux (middle) and detector (bottom) uncertainties. The impact of these systematic uncertainties on the PRISM (left) and On-Axis (right) contours are to be compared 152
Fig. 6.25 Improvement in the precision of PRISM Δm^2_{32} and $\sin^2\theta_{23}$ measurements with increasing FD exposure. Precision is quantified by the width of the 1σ sensitivity band from a single-parameter PRISM fit. A smaller 1σ sensitivity band corresponds to a more precise measurement. The staging plan described in Sect. 6.1 is assumed 153
Fig. 7.1 PRISM predictions for the four DUNE signal channels made in the variable $E^{rec.}_{dep.}$. The total stacked histogram is the PRISM prediction. The methodology used to produce the predictions is detailed in Chap. 4 157
Fig. 7.2 Event rate predictions for ν_μ according to the nominal MC and mock data for the FD (top) and at different off-axis positions at the ND (bottom) in FHC mode. The ND event rates are integrated over the off-axis range indicated in the corresponding legend. The small stacked histograms are the ND NC and wrong-sign backgrounds. There is a minimum at approximately 1 GeV in the ND event rates, which is due to the exclusion of events where the muon is neither contained nor propagates to TMS/ND-GAr 159
Fig. 7.3 Event rate predictions for $\bar{\nu}_\mu$ according to the nominal MC and mock data for the FD (top) and at different off-axis positions at the ND (bottom) in RHC mode. The selected ND event rates are integrated over the off-axis range indicated in the corresponding legend. The small stacked histograms are the ND NC and wrong-sign backgrounds. There is a minimum at approximately 1 GeV in the ND event rates, which is due to the exclusion of events where the muon is neither contained nor propagates to TMS/ND-GAr 160

Fig. 7.4 Contours for fits of the nominal FD MC to the nominal FD MC (dashed contours) and to the mock data (solid contours). A joint fit at the FD and on-axis ND is performed that accounts for all sources of systematic uncertainty. The fit at the on-axis ND constrains the systematic uncertainty parameters. The true oscillation point is indicated by the star and 90% contours for three exposures are shown. Figure taken from Ref. [1] 161
Fig. 7.5 Four signal channel fits of the PRISM (red) and FD MC (blue) predictions to the FD mock data in a Δm^2_{32} versus $\sin^2\theta_{23}$ parameter space. 1σ and 3σ contours are plotted for a single exposure of 100 kt-MW-Yrs 162
Fig. 7.6 FD mock data and PRISM predictions produced by linearly combining ND mock data for the channels $\nu_\mu \rightarrow \nu_\mu$ (top left), $\bar{\nu}_\mu \rightarrow \bar{\nu}_\mu$ (top right), $\nu_\mu \rightarrow \nu_e$ (bottom left) and $\bar{\nu}_\mu \rightarrow \bar{\nu}_e$ (bottom right). There is poor agreement between the PRISM predictions and FD mock data in both channels 162
Fig. 7.7 Four signal channel fit of the PRISM prediction to the FD mock data in a Δm^2_{32} versus $\sin^2\theta_{23}$ parameter space. The mock data proton energy shift and reweighting scheme has been applied to the ND efficiency calculation for the PRISM prediction. 1σ and 3σ contours are plotted for a single exposure of 100 kt-MW-Yrs 163
Fig. 7.8 FHC (left) and RHC (right) signal channel fits of the PRISM predictions to the FD mock data in a Δm^2_{32} versus $\sin^2\theta_{23}$ parameter space. The mock data reweighting scheme has been applied to the ND efficiency calculation for the PRISM prediction. 1σ and 3σ contours are plotted for a single exposure of 100 kt-MW-Yrs 164
Fig. 7.9 FHC (left) and RHC (right) signal channel fits of the PRISM predictions to the FD mock data in a Δm^2_{32} versus $\sin^2\theta_{23}$ parameter space. The mock data reweighting scheme has been applied to the ND efficiency calculation and the FD wrong-sign background for the PRISM prediction. 1σ and 3σ contours are plotted for a single exposure of 100 kt-MW-Yrs 165
Fig. 7.10 Four signal channel fit of the PRISM prediction to the FD mock data in a Δm^2_{32} versus $\sin^2\theta_{23}$ parameter space. The mock data proton energy shift and reweighting scheme has been applied to the ND efficiency calculation and the FD wrong-sign background for the PRISM prediction. 1σ and 3σ contours are plotted for a single exposure of 100 kt-MW-Yrs 166

Fig. 7.11 FD mock data and PRISM predictions produced by linearly combining ND mock data for the channels $\nu_\mu \rightarrow \nu_\mu$ (top left), $\bar{\nu}_\mu \rightarrow \bar{\nu}_\mu$ (top right), $\nu_\mu \rightarrow \nu_e$ (bottom left) and $\bar{\nu}_\mu \rightarrow \bar{\nu}_e$ (bottom right). The mock data proton energy shift and reweighting procedure has now been applied to the ND efficiency calculation and wrong-sign background components of the PRISM predictions. Good agreement has been recovered between the PRISM predictions and FD mock data in all channels 166

Fig. 7.12 Four signal channel fits of the PRISM predictions to the FD mock data for Δm^2_{32}, $\sin^2\theta_{23}$ and δ_{CP}. All flux, cross section and detector systematic uncertainties are accounted for in the fits. The mock data reweighting scheme has been applied to the ND efficiency calculation and the FD wrong-sign background for the PRISM predictions 168

Fig. 7.13 PRISM oscillation contours at a fixed exposure in a two-dimensional oscillation parameter spaces. The reactor neutrino θ_{13} constraint is removed for the $\sin^2 2\theta_{13}$ versus δ_{CP} sensitivity. All flux, cross section and detector systematic uncertainties are accounted for in the fits. The mock data reweighting scheme has been applied to the ND efficiency calculation and the FD wrong-sign background for the PRISM predictions 169

Fig. A.1 A series of "L-Curves" for each of the oscillation channels. The "kink" in the curve corresponds to the optimum regularisation parameter 176

Fig. A.2 Curvature plotted as a function of the regularisation parameter for each oscillation channel. A clear maximum in the curvature can be seen for each oscillation channel, indicating the optimum value for τ_{LC} 177

List of Tables

Table 2.1 Best fit values and uncertainties of the three-flavour oscillation model parameters for normal ordering ($\Delta m^2_{32} > 0$) and inverted ordering ($\Delta m^2_{32} < 0$). Results are from References [76] and [77] 21

Table 3.1 The POT generated for the ND MC data at each off-axis stop in FHC mode and RHC mode. One year of data taking corresponds to 1.1×10^{21} POT. Therefore, in total there is 6.58 years of FHC ND MC and 7.73 years of RHC ND MC assuming a 1.2 MW beam power 54

Table 5.1 Uncertainties on the flux model parameters. The same set of systematic uncertainty parameters affects the flux at the ND and FD. Parameters are categorised based on the component of the flux model they impact 95

Table 5.2 Uncertainties on the interaction model parameters in the GENIE generator. The variable W refers to the energy transferred to the nucleus in a neutrino interaction. Events with a low-W that are not CCQE tend to be resonance production events. High-W events tend to be DIS. The initials "BY" refer to the Bodek-Yang model of DIS [10] 105

Table 5.3 Additional cross section uncertainty parameters not contained in GENIE. There are three BeRPA parameters covering different regions of Q^2 space and 23 DIS parameters for CC/NC neutrino/antineutrino scattering on protons/neutrons with one to three pions in the final state. Separate 2p-2h parameters are implemented for neutrino and antineutrino scattering 106

Table 5.4 Uncertainties on the energy scale parameters. The same parameters are used for the ND and FD energy scale uncertainties. Correlated uncertainties are 100% correlated between the ND and FD. Uncorrelated uncertainties are totally uncorrelated 112
Table 6.1 Mapping from years of exposure to kt-MW-Yrs. The mapping accounts for the assumed FD staging plan for module installation and the PIP-II upgrade 120
Table 6.2 Best fit values and uncertainties of the oscillation parameters. Normal mass ordering is assumed. Certain parameters have their 1σ error approximated to a Gaussian uncertainty. The final column indicates whether constraints derived from the parameter uncertainty are applied to the parameter in the oscillation fit. The oscillation parameters and their uncertainties are from the 2018 NuFIT 4.0 results [2] and the Earth's density is taken from Ref. [3] 121
Table 6.3 Summary of the single-parameter sensitivities from Figs. 6.5 and 6.16 with the MC-based (left column) and data-driven (right column) ND efficiency correction respectively. The uncertainties refer the 1σ parameter uncertainty over a FD exposure of 1056 kt-MW-Yrs 145
Table 6.4 Mapping from years of exposure to kt-MW-Yrs according to Refs. [1, 7]. The mapping accounts for the assumed FD staging plan and the PIP-II upgrade 147
Table A.1 Optimum τ_{LC} parameters for each of the signal channels. These are the parameters used for the PRISM analysis presented in this thesis 177

Chapter 1
Introduction

The leading theory for the development of the early universe asserts that the universe began in a Big Bang in which equal quantities of matter and antimatter were created. However, we know that our observable universe is entirely dominated by matter. How this came to be the case is one of the most persistent problems in cosmology and particle physics. The necessary conditions to create a matter-antimatter asymmetry in the early universe are well known; these are the Sakharov conditions [1]. The difficulty is in understanding the mechanisms through which these conditions are met. One of the conditions is the existence of particle interactions that violate CP symmetry, which is observed as different physics for matter and antimatter particles. If the Standard Model of particle physics were to conserve CP symmetry and therefore not distinguish between matter and antimatter interactions, then our matter-dominated universe would simply not exist. Consequently, incorporating sufficient CP symmetry violation into the Standard Model would constitute a major step towards understanding the origin of the matter-antimatter asymmetry in the universe.

Violation of CP symmetry in quark mixing has been studied extensively. However, the CP violation observed in the quark sector is not sufficient to resolve the matter-antimatter asymmetry problem. It is possible that the mixing of neutrino flavours could provide an additional source of CP symmetry violation that could resolve this mystery. Neutrino flavour mixing occurs through the phenomenon of neutrino oscillations. An accessible overview of neutrinos and their potential role in explaining the matter-antimatter asymmetry can be found in [2], whilst more technical and mathematical treatments are given in [3], and Chapter 11 of [4]. Nonetheless, definitive evidence of CP symmetry violation in neutrino oscillations through a measurement of the CP-violating phase δ_{CP} has yet to be observed. This has led to the development of a new generation of experiments that aim to conclusively determine whether the violation of CP symmetry occurs in neutrino oscillations.

C. M. Hasnip, *DUNE-PRISM—A New Method to Measure Neutrino Oscillations*,
Springer Theses, https://doi.org/10.1007/978-3-032-07251-1_1

The Deep Underground Neutrino Experiment (DUNE) is a long baseline neutrino oscillation experiment that has been designed to discover CP symmetry violation in the lepton sector, determine the neutrino mass ordering and precisely measure the parameters Δm^2_{32}, θ_{23}, θ_{13} and δ_{CP}. To do this, DUNE employs the world's most intense neutrino beam, directed along a 1285 km baseline to a series of massive neutrino detectors. At the time of writing, DUNE is under construction and will provide world-leading measurements of neutrino oscillations in the years to come.

The power of the DUNE neutrino beam and the scale of the detectors mean that DUNE will not be limited by the size of sample it can collect. Instead, DUNE's performance will be determined by the size of the systematic uncertainties. One of the most problematic sources of systematic uncertainty is the neutrino interaction model, which is relied upon to relate the final state particles observed in a neutrino interaction to the initial neutrino energy. Large systematic uncertainties are associated with the many parameters of currently available neutrino interaction models. Worse, no model is likely to exist on the timescale of DUNE that fully describes the physics of neutrino interactions with large nuclei, leading to the inclusion of empirical corrections that force agreement between the model predictions and the data. A poor choice of alterations to the neutrino interaction model can bias the relationship between the true neutrino energy and the observable energy, causing a bias in the measurement of the oscillation parameters.

This thesis aims to address the challenges associated with modelling neutrino interactions through a new oscillation analysis methodology called the DUNE Precision Reaction Independent Spectrum Measurement (DUNE-PRISM). The DUNE-PRISM (shortened to just "PRISM" for conciseness) method employs a movable detector to measure different neutrino fluxes that can be linearly combined to produce a data-driven prediction of the oscillated neutrino event rate. An oscillated event rate prediction that is primarily composed of data naturally contains the correct neutrino interaction physics, rather than relying on a potentially inaccurate model. This thesis will be the first to demonstrate how the PRISM method can be used to make a neutrino oscillation measurement for DUNE. In addition, it is hoped that this thesis will prove to be a useful guide to any current or future neutrino physicist interested in further developing the data-driven neutrino oscillation analysis presented here.

An overview of neutrino physics is provided in Chap. 2. This overview covers a brief history of neutrino physics, the physics of neutrino interactions in the Standard Model and the theory and experimental status of neutrino oscillations. In addition, Chap. 2 provides a discussion on the challenges involved in modelling neutrino interactions, which, as mentioned above, is the primary motivation for developing a new data-driven oscillation analysis.

DUNE is a large, complex neutrino oscillation experiment and is composed of multiple detectors and facilities across two separate sites. Chap. 3 describes each component of DUNE in detail. This includes the production of the neutrino beam, the near detector and the far detector. The chapter ends with a description of the simulated data, produced by members of the DUNE Collaboration, that are used to produce all the results presented in this thesis.

Chapter 4 provides a step-by-step guide to the PRISM methodology. Producing a PRISM prediction of the oscillated event rate is a complicated procedure and each step of the process is explained in detail. The chapter demonstrates how a PRISM prediction can be made for any oscillation hypothesis, from which the oscillation parameters can be measured. This chapter references Appendix A, which provides technical details on an optimisation of the linear combination procedure. As this is the first thesis on the PRISM oscillation analysis, there are some features of the methodology that are still under development. Chap. 4 therefore finishes with a discussion on the planned improvements to the methodology for the next iteration of the PRISM analysis.

As mentioned above, understanding the sources and magnitude of the systematic uncertainties is of great importance for DUNE. Chapter 5 explains how the different systematic uncertainties impact the PRISM oscillation analysis. This includes a series of studies that quantify the expected impact of each category of systematic uncertainty on the oscillation measurement. The most important sources of systematic uncertainty are identified through these studies.

The PRISM methodology is then applied to a demonstration of a full, realistic PRISM measurement of the oscillation parameters in Chap. 6. Predicted PRISM oscillation sensitivities are shown that account for a full systematic uncertainty model. The relative impact of the flux, cross section and detector uncertainties on the PRISM sensitivities are then studied and the most important systematic uncertainties are singled out for each of the oscillation parameter measurements. Further sensitivities are shown that simulate the inclusion of one of the methodology improvements described in Chap. 4. Finally, comparisons are made between the PRISM sensitivities and the published DUNE sensitivities produced using a traditional long baseline oscillation analysis method [5]. The final section of Chap. 6 discusses the source of any differences in sensitivity between the two analyses.

The possibility of obtaining a biased oscillation parameter measurement due to errors in the neutrino interaction model motivated the development of the PRISM oscillation analysis. Chap. 7 explores a scenario in which a poor choice of alterations to the neutrino interaction model has indeed resulted in a biased relationship between the true neutrino energy and the observable energy of the neutrino interaction. It is demonstrated that, whilst the traditional DUNE oscillation analysis gets the measurement of the oscillation parameters wrong, the PRISM analysis obtains a measurement free from bias.

This thesis concludes with Chap. 8, in which the key results and future plans for the PRISM oscillation analysis are discussed. Finally, a road-map is established that proposes the next steps to be taken in further improving the PRISM oscillation analysis.

References

1. A.D. Sakharov, Violation of cp invariance, c asymmetry, and baryon asymmetry of the universe. Soviet Phys. Uspekhi **34**(5), 392 (1991)
2. F. Close, Neutrino (Oxford University Press, Oxford, UK, 2010)
3. B. Kayser, in *Neutrino physics*, eConf, vol. C040802, ed. by J. Hewett, J. Jaros, T. Kamae, C. Prescott, p. L004 (2004). arXiv: hep-ph/0506165
4. M. Fukugita, T. Yanagida, Baryon asymmetry in the universe and neutrinos, in *Physics of Neutrinos: and Application to Astrophysics* (Springer, Berlin Heidelberg, 2003), pp. 487–516
5. DUNE Collaboration, B. Abi, et al., Deep Underground Neutrino Experiment (DUNE), Far Detector Technical Design Report, Volume II: DUNE Physics, Feb. 2020. arXiv: 2002.03005 [hep-ex]

Chapter 2
Neutrino Theory and Background

This chapter provides an overview of the field of neutrino physics, where particular focus is given to the theory and experimental status of neutrino oscillations. The chapter begins with a discussion on the history of neutrino physics from its earliest days. Accounts of this history that are aimed at a general audience can be found in [1, 2]. An overview of the theory of neutrinos interactions in the Standard Model is then followed by a review of the phenomena of neutrino oscillations. Textbooks by Griffiths [3], Martin and Shaw [4] and Thomson [5] all provide excellent introductions to the Standard Model of particle physics aimed at physics undergraduates and post-graduates. More in-depth theoretical treatments of neutrino masses and oscillations that are still highly accessible can be found in the lectures of Boris Kayser [6, 7]. The contribution of past, present and future neutrino oscillation experiments to the field are then addressed. Cahn and Goldhaber's textbook provides an excellent overview of many different past neutrino experiments with references and figures from the original papers [8]. The chapter ends with an extended discussion on the challenges involved in accurately modelling neutrino interactions with matter, which is of particular relevance to this thesis.

2.1 Neutrinos: A Brief History

In 1914 James Chadwick observed a continuous electron energy spectrum in nuclear β decay—a discovery which confounded theoretical expectations based on energy conservation [9]. If the β radiation was a two-body nuclear decay, as was believed, then the emitted electron should have been monoenergetic. Previous experiments investigating nuclear α and γ radiation had observed discrete energy spectra consistent with the energy difference between the initial and final state nucleus. The theoretical conundrum drove Wolfgang Pauli to propose a "desperate remedy" [10],

C. M. Hasnip, *DUNE-PRISM—A New Method to Measure Neutrino Oscillations*,
Springer Theses, https://doi.org/10.1007/978-3-032-07251-1_2

where the continuous electron energy spectrum in β decay could be explained by an unobserved spin-1/2 neutral particle with a small mass he called a *neutron*. The *neutron* would be emitted in β decay in addition to the electron, carrying away some of the energy and hence preserving energy and angular momentum conservation. The *neutron* was soon renamed the *neutrino* by Enrico Fermi and what we now call the neutron was discovered by Chadwick in 1932 [11]. A particle of low mass and neutral charge would be challenging to detect, with Pauli himself questioning whether detection was even possible. Using Fermi's theory of the weak interaction [12], Bethe and Peierls calculated the cross section of a neutrino interacting with a proton to be smaller than 10^{-44} cm^2, concluding, understandably, that such an interaction would be impossible to observe [13].

Fortunately, Bethe and Peierls' conclusions were made before the advent of the nuclear power and improvements in detector technology. By the mid-1950s, detection of the elusive neutrino had been achieved by Reines and Cowan of Los Alamos Laboratory [14, 15]. A liquid scintillator detector loaded with a cadmium compound was positioned close to the Savannah River nuclear reactor. The experiment employed a delayed-coincidence technique to first measure the positron annihilation from the inverse β decay process,

$$\bar{\nu} + p \rightarrow e^+ + n, \tag{2.1}$$

followed by the capture of the neutron on cadmium, which resulted in the emission of γ rays. A second scintillator detector without the cadmium loading was placed above the first to reject cosmic ray background events.

As Cowan and Reines collected neutrino interactions at Savannah River, Goldhaber, Grodzins and Sunkar measured the circular polarisation of resonant scattered γ rays emitted shortly after the electron capture decay process

$$e^- + {}^{152\,\mathrm{m}}\mathrm{Eu} \rightarrow {}^{152*}\mathrm{Sm} + \nu. \tag{2.2}$$

These measurements determined the neutrino to be exclusively left-handed, or negative helicity [16]. The left-handed nature of the neutrino also supported the V-A theory of the weak interaction developed by Marshak and Sudarshan and by Feynman and Gell-Mann [17, 18].

The possibility of measuring the interactions of neutrinos produced by accelerator beams through the decays $\pi \rightarrow \mu\nu$ and $K \rightarrow \mu\nu$ was proposed independently by Bruno Pontecorvo and Melvin Schwartz [19, 20]. A team at Brookhaven Laboratory first achieved these measurements in 1962 using the Alternating Gradient Synchrotron (AGS). The AGS directed a 15 GeV proton beam on to a beryllium target, producing a large flux of pions and kaons, which subsequently decayed into neutrinos. The detector was a 10 ton spark chamber. The experiment observed significantly more muon events than electron events, demonstrating that there were at least two neutrino flavours, ν_e and ν_μ [21].

The discovery of the τ lepton at the SPEAR collider at Stanford Linear Accelerator Center (SLAC) in 1975 indicated the possible existence of a third flavour of neutrino [22]. The ν_τ was first observed in 2001 by the DONUT experiment at Fermilab, which directed a neutrino beam at a nuclear emulsion target. The ν_τ in the beam interacted in the emulsion to produce a τ that subsequently decayed, exhibiting a characteristic "kink" in the decaying τ track [23].

Measurements of electron neutrinos from the Sun (solar neutrinos) were made from the 1960s onward, beginning at the Homestake Mine chlorine experiment pioneered by Ray Davis [24]. The Homestake experiment measured neutrinos from ^{8}B decays in the Sun capturing on ^{37}Cl. The resulting ^{37}Ar atoms were counted to give the solar neutrino event rate. Davis consistently measured a deficit in the rate of solar neutrinos as compared to the rate predicted by the Standard Solar Model (SSM) [25]. The so-called "solar neutrino problem" persisted into the 1990s, when the radiochemical GALLEX detector, which measured the capture of solar neutrinos on ^{71}Ga, observed a solar neutrino rate roughly 60–70% of the predicted rate [26]. Similar results were found by the SAGE detector [27]. A reliable model for the solar neutrino flux and a range of consistent experimental results using different methodologies left the possibility that the behaviour of the neutrinos could be the source of the solar neutrino deficit.

Bruno Pontecorvo first proposed that lepton number could be violated through neutrino-antineutrino transitions [28, 29]. Subsequently, following the discovery of two distinct neutrino flavours, ν_e and ν_μ, Pontecorvo discussed whether a non-zero neutrino mass and the violation of lepton number could lead to mixing between different neutrino flavours [30]. Discrepancies between measured and predicted solar neutrino fluxes could therefore be explained if some neutrinos transitioned to a different, unmeasured, neutrino flavour during propagation. This theory was confirmed in the early 2000s by the Super-Kamiokande, SNO and KamLAND experiments [31–33], demonstrating that neutrinos are massive leptons that oscillate between the three lepton flavours: ν_e, ν_μ and ν_τ. The rest of this chapter will further discuss the theory and measurements of neutrino interactions and oscillations.

2.2 Neutrinos in the Standard Model

Our current best understanding of the properties and interactions of matter is contained in the Standard Model (SM). At its most fundamental, all matter can be described by 12 spin-1/2 fermions: 6 quarks and 6 leptons, each categorised into three generations. The three quark generations are

$$\begin{pmatrix} u \\ d \end{pmatrix} \begin{pmatrix} c \\ s \end{pmatrix} \begin{pmatrix} t \\ b \end{pmatrix} \tag{2.3}$$

and the three lepton generations are

$$\begin{pmatrix} \nu_e \\ e \end{pmatrix} \begin{pmatrix} \nu_\mu \\ \mu \end{pmatrix} \begin{pmatrix} \nu_\tau \\ \tau \end{pmatrix}, \tag{2.4}$$

which correspond to the three weak-interaction states: electron, muon and tau. The u, c and t quarks have charge $+2/3$ and the d, s and b quarks have charge $-1/3$. Charged leptons (e, μ and τ) have charge -1 and the neutrinos carry no charge. The sign of the electric charge is reversed for the corresponding antiparticle. Charged particles can interact via the electromagnetic force. In addition to electric charge, quarks carry colour charge and interact through gluons; the carrier of the strong force. The neutrinos have no electric or colour charge and so only interact weakly.

2.2.1 The Weak Interaction

The Fermi theory of the weak interactions proved highly successful at predicting experimental results for many decades. However, the "four-point interaction" violated the unitarity principle by predicting the neutrino scattering cross section to grow linearly with neutrino energy. In addition, the original formalism of the Fermi theory of weak interactions failed to accommodate a mechanism for parity violation, which was first discovered by Wu in 1957 by measuring the emission of β decay electrons from ^{60}Co nuclei [34].

Parity violation was incorporated into the weak interaction through V-A ("vector minus axial vector") theory. When calculating the matrix element of a weak interaction process, each incoming and outgoing fermion, represented by a Dirac spinor u, contributes a current

$$j_{\mu} \propto \bar{u}\gamma_{\mu}(1 - \gamma_5)u, \tag{2.5}$$

where γ_{μ} are the Dirac matrices and $\gamma_5 = i\gamma_0\gamma_1\gamma_2\gamma_3$. Whilst the vector component, γ_{μ}, and axial vector component, $\gamma_{\mu}\gamma_5$, individually conserve parity, the interference between the two components maximally violates parity. It is also due to the V-A structure of the weak interaction that only left-handed helicity neutrinos and right-handed helicity antineutrinos have been experimentally observed.

The non-unitarity of Fermi theory was resolved by the Standard Model of electroweak interactions, developed by Glashow, Weinberg and Salam [35–37]. Electroweak theory predicts a massless photon, the mediator of the electromagnetic force, and three massive bosons: W^+, W^- and Z^0. Hence, the existence of the weak neutral current (NC), mediated by the Z^0, in addition to the weak charged current (CC) was a key prediction of electroweak theory. The masses of the gauge bosons are acquired through the Higgs mechanism [38–40]. Mediating the weak force via massive bosons (two charged and one neutral) ensures that the weak interaction is both short ranged and preserves unitarity. Figure 2.1 shows Feynman diagrams for CC and NC neutrino scattering off a nucleon. The weak NC was discovered at the Gargamelle detector in CERN and the $W^{\pm}$ and Z^0 were discovered in the 1980s by the UA1 and UA2 collaborations [41–43].

Following the discovery of the weak bosons, the width of the Z^0 resonance at the Large Electron-Positron (LEP) collider was used to determine the number of neutrino flavours that couple to the weak force. The width of any resonance is partly

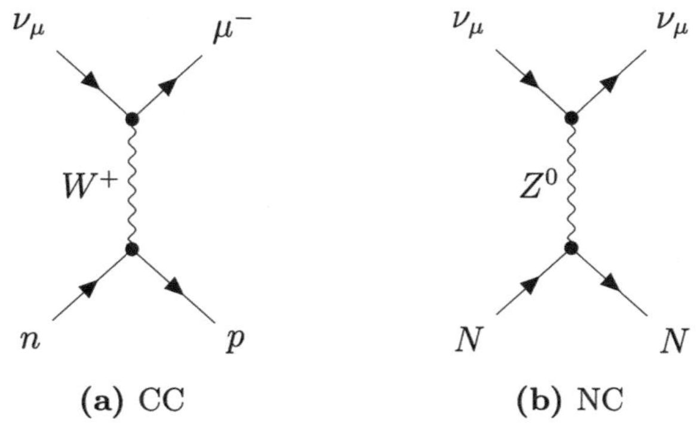

Fig. 2.1 Charged current (CC) and neutral current (NC) scattering of a neutrino off a nucleon. Here, $N = p, n$

determined by the number of channels the resonance can decay to. The LEP study found that the width of the Z^0 was consistent with exactly three species of neutrino that couple to the weak force and have a mass less than half the Z^0 mass [44], as can be seen in Fig. 2.2.

2.3 Neutrino Oscillations

From studying the weak interaction between quarks in hadronic states, it is known that particle mass states are not necessarily the same as the weak interaction states. This allows for the mixing between different generations of quarks during hadronic weak interactions. The relationship between the mass and weak eigenstates of the quarks is captured by the Cabibbo-Kobayashi-Maskawa (CKM) matrix [45, 46]. If neutrinos have a small, but non-zero, mass, a similar phenomena may occur for neutrino states. The neutrino weak eigenstates ν_e, ν_μ and ν_τ and mass eigenstates ν_1, ν_2 and ν_3 can be related to each other by a CKM-like matrix:

$$\begin{pmatrix} \nu_e \\ \nu_\mu \\ \nu_\tau \end{pmatrix} = \begin{pmatrix} U_{e1} & U_{e2} & U_{e3} \\ U_{\mu 1} & U_{\mu 2} & U_{\mu 3} \\ U_{\tau 1} & U_{\tau 2} & U_{\tau 3} \end{pmatrix} \begin{pmatrix} \nu_1 \\ \nu_2 \\ \nu_3 \end{pmatrix}, \tag{2.6}$$

which is known as the Pontecorvo-Maki-Nakagawa-Sakata (PMNS) matrix [29, 47]. An electron neutrino, produced in a CC weak interaction in conjunction with a positron, therefore propagates as the linear combination of the mass eigenstates:

$$|\psi\rangle = U_{e1}\,|\nu_1\rangle + U_{e2}\,|\nu_2\rangle + U_{e3}\,|\nu_3\rangle\,. \tag{2.7}$$

Upon a CC interaction, the wavefunction collapses back to a weak eigenstate, producing a charged lepton of a particular flavour. If the masses associated with the three neutrino mass eigenstates are different, then phase differences develop between the

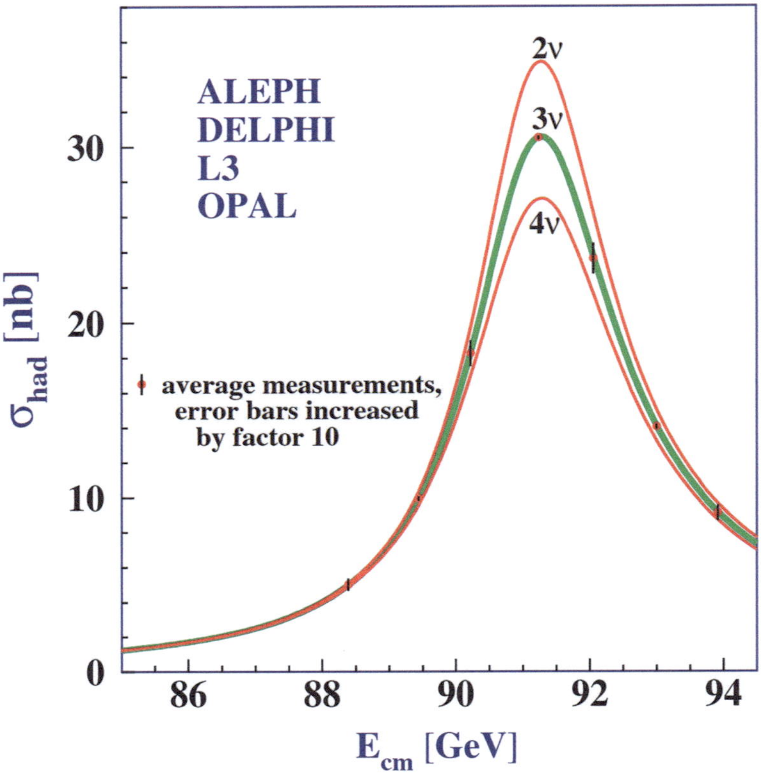

Fig. 2.2 The resonance of the Z^0 as measured by LEP. The width is consistent with exactly three flavours of weakly interacting neutrinos. Reprinted from Phys. Rep., vol. 427 (5—6), The ALEPH Collaboration et al., "Precision Electroweak Measurements on the Z Resonance", pp. 257—454, Copyright (2006), with permission from Elsevier [44]

different contributions to the neutrino wavefunction. This allows for a certain probability that the lepton flavour observed upon interaction is different from the lepton flavour associated with the production of the neutrino. Such a phenomena is known as neutrino oscillations. This section will explain how a non-zero neutrino mass leads to the oscillation of neutrino flavour, beginning with the two-flavour approximation.

2.3.1 Neutrino Oscillations of Two Flavours

The main features of neutrino oscillations may be captured by considering two neutrino flavours, ν_α and ν_β. In the two-flavour case the weak and mass eigenstates are related by a 2×2 unitary matrix characterised by a mixing angle θ,

$$\begin{pmatrix} \nu_\alpha \\ \nu_\beta \end{pmatrix} = \begin{pmatrix} \cos\theta & \sin\theta \\ -\sin\theta & \cos\theta \end{pmatrix} \begin{pmatrix} \nu_1 \\ \nu_2 \end{pmatrix}. \tag{2.8}$$

Hence the wavefunction ν_α, produced as a linear superposition of ν_1 and ν_2, evolves with time, t, in position-space, $\mathbf{x}$, according to

$$|\psi(\mathbf{x},t)\rangle = \cos\theta\,|\nu_1\rangle\,e^{-i\phi_1} + \sin\theta\,|\nu_2\rangle\,e^{-i\phi_2}, \tag{2.9}$$

where $\phi_i = E_i t - \mathbf{p_i x}$ is a phase factor that is expressed in terms of the energy, E_i, and three-momentum, $\mathbf{p_i}$, of the ν_i mass state. Equation 2.9 can be rewritten as a linear superposition of weak eigenstates by taking the inverse of Eq. 2.8. The probability of a neutrino initially produced in the ν_α state to oscillate to the ν_β state upon interaction after propagating a given distance is therefore

$$P(\nu_\alpha \rightarrow \nu_\beta) = |\langle\nu_\beta|\,|\psi(\mathbf{x},t)\rangle\,|^2 = \sin^2(2\theta)\sin^2\left(\frac{\phi_2-\phi_1}{2}\right). \tag{2.10}$$

By assuming the neutrino mass eigenstates are produced with the same energy, E, and that the neutrino travels at relativistic speeds over a distance $L = |\mathbf{x}|$, the difference in phase factors can be approximated to be

$$\phi_2 - \phi_1 \approx \frac{(m_2^2 - m_1^2)L}{2E}, \tag{2.11}$$

where m_1^2 and m_2^2 are the squared masses of the ν_1 and ν_2 mass eigenstates respectively. Substituting the approximate phase difference into Eq. 2.10 yields the two flavour oscillation probability

$$P(\nu_\alpha \rightarrow \nu_\beta) = \sin^2(2\theta)\sin^2\left(\frac{(m_2^2 - m_1^2)L}{4E}\right). \tag{2.12}$$

It is common to write Eq. 2.12 in terms of the units suitable to the length and energy scales in which neutrino oscillations have been observed. Expressing energy in GeV, the mass-squared difference in eV^2 and length in kilometres, Eq. 2.12 can expressed as

$$P(\nu_\alpha \rightarrow \nu_\beta) = \sin^2(2\theta)\sin^2\left(1.27\frac{\Delta m_{21}^2[\text{eV}^2]L[\text{km}]}{E[\text{GeV}]}\right), \tag{2.13}$$

where $\Delta m_{21}^2 = m_2^2 - m_1^2$. The probability of having the same neutrino flavour at interaction as at production – known as the survival probability – is simply given by $P(\nu_\alpha \rightarrow \nu_\alpha) = 1 - P(\nu_\alpha \rightarrow \nu_\beta)$.

2.3.2 Neutrino Oscillations of Three Flavours

Oscillation probabilities can be derived for any number of neutrino flavours using the same method employed to calculate the two-flavour oscillation probability. However, this section will limit the discussion to the three neutrino flavours contained in the Standard Model. The weak eigenstates are related to the mass eigenstates by the same 3×3 unitary PMNS matrix shown in Eq. 2.6. As in the two-flavour case, a neutrino, produced in the weak eigenstate ν_μ, propagates as a linear superposition of the three mass eigenstates, ν_k, and at some later t the wavefunction is

$$|\psi(\mathbf{x}, t)\rangle = \sum_k U_{\mu k} e^{-i\phi_k} |\nu_k\rangle . \tag{2.14}$$

After rewriting Eq. 2.14 in terms of the weak eigenstates, the probability of observing a neutrino of flavour ν_e after propagating a distance L is

$$\begin{aligned} P(\nu_\mu \to \nu_e) &= |\langle \nu_e| \, |\psi(\mathbf{x}, t)\rangle|^2 \\ &= \sum_{k,j} U_{\mu k} U^*_{ek} U^*_{\mu j} U_{ej} e^{-i(\phi_k - \phi_j)}. \end{aligned} \tag{2.15}$$

Using the same approximations for the phase differences as Sect. 2.3.1, the oscillation probability can be expressed as

$$\begin{aligned} P(\nu_\mu \to \nu_e) = \delta_{\mu e} &- 4 \sum_{k>j} Re\left(U_{\mu k} U^*_{ek} U^*_{\mu j} U_{ej}\right) \sin^2\left(\frac{\Delta m^2_{kj} L}{4E}\right) \\ &+ (-)2 \sum_{k>j} Im\left(U_{\mu k} U^*_{ek} U^*_{\mu j} U_{ej}\right) \sin\left(\frac{\Delta m^2_{kj} L}{4E}\right), \end{aligned} \tag{2.16}$$

where $\delta_{\mu e}$ is the Kronecker delta function. Like in Eq. 2.13, Δm^2_{kj} is the mass-squared difference between mass eigenstates ν_k and ν_j, where the allowed combinations are Δm^2_{21} and Δm^2_{32}.

The sign on the imaginary component of Eq. 2.16 becomes negative in case of $\bar{\nu}_\mu \to \bar{\nu}_e$ antineutrino oscillations, which is the CP conjugate process to $\nu_\mu \to \nu_e$ oscillations. Therefore, if the PMNS matrix elements are complex $P(\nu_\mu \to \nu_e) \neq P(\bar{\nu}_\mu \to \bar{\nu}_e)$. This raises the possibility that CP symmetry is violated in three-flavour neutrino oscillations.

In three-flavour oscillations the PMNS matrix is typically expressed in terms of three mixing angles (θ_{12}, θ_{23} and θ_{13}), a CP-violating phase δ_{CP} and two Majorana phases $\alpha_{1,2}$:

$$\begin{aligned} U \equiv &\begin{pmatrix} 1 & 0 & 0 \\ 0 & c_{23} & s_{23} \\ 0 & -s_{23} & c_{23} \end{pmatrix} \begin{pmatrix} c_{13} & 0 & s_{13}e^{-i\delta_{CP}} \\ 0 & 1 & 0 \\ -s_{13}e^{i\delta_{CP}} & 0 & c_{13} \end{pmatrix} \\ &\times \begin{pmatrix} c_{12} & s_{12} & 0 \\ -s_{12} & c_{12} & 0 \\ 0 & 0 & 1 \end{pmatrix} \begin{pmatrix} e^{i\alpha_1/2} & 0 & 0 \\ 0 & e^{i\alpha_1/2} & 0 \\ 0 & 0 & 1 \end{pmatrix}, \end{aligned} \tag{2.17}$$

where $s_{ij} = \sin\theta_{ij}$ and $c_{ij} = \cos\theta_{ij}$. Neutrino and antineutrino oscillation probabilities are only different if the PMNS matrix has complex elements. Therefore, CP violation occurs in neutrino oscillations if δ_{CP} is neither zero nor an integer value of π. In addition, it can be shown that CP violation in neutrino oscillations only occurs if there is mixing between at least three flavours, requiring all three mixing angles to be non-zero [48]. The Majorana phases are only relevant if the neutrinos are Majorana particles and the values of the Majorana phases do not affect the oscillation probability calculation. Hence, whether neutrinos are Dirac of Majorana particles cannot be experimentally determined by studying neutrino oscillations.

2.3.3 Neutrino Oscillations in Matter

The discussion on neutrino oscillations has so far focused on oscillations in a vacuum, however propagation through matter can significantly alter the neutrino oscillation probability through the Mikheyev-Smirnov-Wolfenstein (MSW) effect [49, 50]. This section will only summarise the key features of the MSW effect, but a comprehensive treatment can be found in [51]. during propagation though matter, any neutrino flavour can interact via a NC interaction, but this is of no significance to neutrino oscillations since the NC scattering amplitude is the same for all neutrino flavours. However, the electron neutrino has an additional CC scattering mode, as shown in Fig. 2.3. This extra interaction changes the scattering potential for the electron neutrinos, which modifies the Hamiltonian of the propagating neutrino.

Working with the two-flavour approximation discussed in Sect. 2.3.1, the mixing angle for neutrinos in the presence of matter of a constant density is altered to be

$$\tan 2\theta \rightarrow \tan 2\theta_m = \frac{\tan 2\theta}{1 - A \sec 2\theta}, \tag{2.18}$$

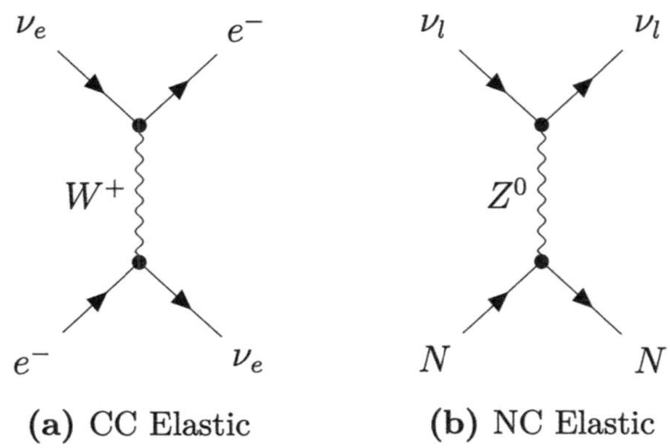

Fig. 2.3 $l = e, \mu$ or τ and $N = e, p$ or n

where $A = 2\sqrt{2}G_F N_e E/\Delta m^2$ and G_F is the Fermi coupling constant. Oscillations in matter are therefore altered according to the electron density, N_e, the neutrino energy, E, and the magnitude *and* sign of Δm^2. If antineutrinos are propagating through matter, then the sign of A is reversed, causing differences between neutrino and antineutrino oscillation probabilities that are not due to the CP violation in neutrino mixing discussed in Sect. 2.3.2. Equation 2.18 shows a resonant condition exists if $A = \cos 2\theta > 0$. The resonant condition is satisfied for neutrino if $\Delta m^2 > 0$ and for antineutrinos if $\Delta m^2 < 0$. Consequently, there can be large neutrino oscillation probabilities in matter even if θ is small.

2.4 Neutrino Oscillation Measurements

So far, this chapter has covered the history of neutrino physics and the key features of the theory of neutrino interactions and oscillations. This section will summarise the current status of neutrino oscillation measurements. There are, in general, four main types of neutrino oscillation experiment: solar neutrinos, atmospheric neutrinos, nuclear reactor neutrinos and accelerator neutrino beams. Each type of experiment probes different neutrino flavour transitions, energies and distances of propagation. Obtaining precise measurements of all the parameters of the PMNS matrix and the squared-mass splittings requires input from all four types of neutrino oscillation experiment.

2.4.1 *Solar Neutrinos*

Observations of neutrinos from the Sun provided the first evidence for neutrino oscillations. As discussed in Sect. 2.1, several radio-chemical experiments (Homestake, GALLEX and SAGE) measured electron neutrino fluxes significantly smaller than the prediction. The first generation of experiments observed CC neutrino interactions at energies below 30 MeV. Greater energy is required to produce the mass of the muon or tau lepton in a CC ν_μ or ν_τ interaction, and therefore only CC ν_e interactions are accessible at these energies. However, NC interactions are possible for all neutrino flavours at all energies and this was the channel pursued by the Sudbury Neutrino Observatory (SNO) experiment.

The SNO detector was a 1000 ton tank of pure heavy water (D_2O) surrounded by 10,000 photomultiplier tubes. Like the original radio-chemical experiments, SNO measured the CC interactions of electron neutrinos from the Sun,

$$\nu_e + D \rightarrow p + p + e^- . \tag{2.19}$$

In addition, for any neutrino flavour ν_l, SNO measured the NC process

$$\nu_l + D \rightarrow p + n + \nu_l \tag{2.20}$$

and the elastic scattering (ES) process

$$\nu_l + e^- \to \nu_l + e^-. \tag{2.21}$$

Although any neutrino flavour can contribute to the elastic scattering process, the cross section is larger for electron neutrinos since it is able to interact through both charged and neutral currents. The first SNO results measured a higher ES flux than CC flux, demonstrating, in conjunction with the Super-Kamiokande ES measurement [52], that muon and tau neutrinos were contributing to the elastic scattering flux through NC interactions [53]. SNO measured the NC channel by observing photon emission from the neutron capturing on deuterium. The total NC flux, to which all three neutrino flavours could contribute equally, was found to agree well with the solar neutrino flux prediction [32]. The SNO experiment therefore confirmed the hypothesis that neutrino flavours mix.

The observations of SNO and other solar neutrino experiments cannot be fully explained in a neutrino flavour mixing model without the inclusion of the MSW effect [53]. One example of this is the electron neutrino component of the total neutrino flux measured at SNO being too small to be explained by only vacuum oscillations [54]. This provides strong evidence that the mixing of neutrino flavour is enhanced by the MSW effect in the Sun. Consequently, the ν_2 state must be heavier than ν_1, which has a larger ν_e flavour component, and $\Delta m^2_{21} > 0$.

Final confirmation that neutrino oscillations resolved the"solar neutrino problem" was provided by the KamLAND experiment. The 1000 ton liquid scintillator detector measured electron antineutrinos from several Japanese nuclear reactors located at distances ranging from 138 km to 214 km from the detector. Significantly, KamLAND was able to measure the electron antineutrino survival probability as a function of L/E, as shown in Fig. 2.4, providing precise measurements of the solar mixing parameters Δm^2_{21} and θ_{12} [55]. The solar neutrino measurements and KamLAND thus demonstrated that the "solar neutrino problem" is best resolved by neutrino oscillations and the MSW effect.

2.4.2 Atmospheric Neutrinos

Before the SNO and KamLAND experiments had confirmed the existence of solar neutrino oscillations, atmospheric neutrinos produced by cosmic rays had provided convincing evidence for the existence of neutrino oscillations. Cosmic rays striking the atmosphere create showers of hadrons, a large proportion of which are pions. The charged pion decay process is $\pi \to \mu\nu_\mu$ and the muon subsequently decays to an electron via $\mu \to e\nu_e\nu_\mu$. Therefore, in the absence of the neutrino oscillations, for every electron neutrino two muon neutrinos are expected at the Earth's surface.

Large-mass water Cherenkov detectors were constructed in the 1980s to search for hypothesised nucleon decay processes. Atmospheric neutrinos with energies at the GeV-scale were studied as a potential source of background to such a search.

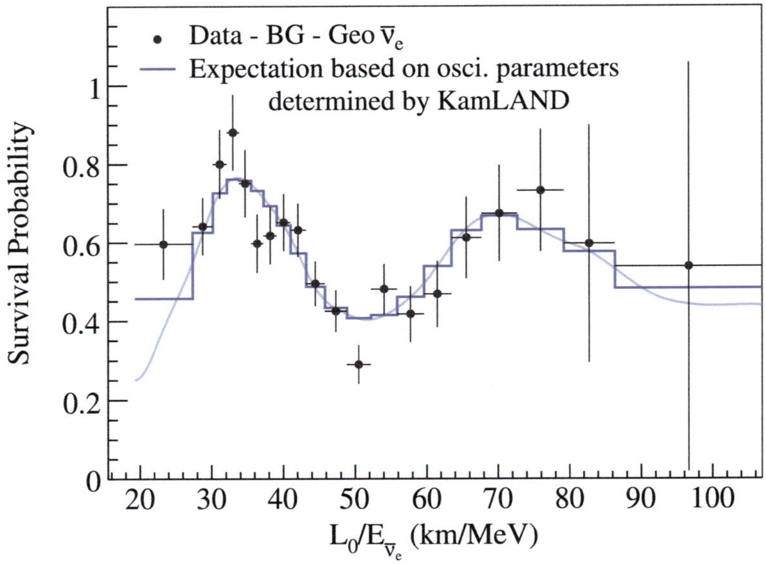

Fig. 2.4 Ratio of the measured background and geoneutrino subtracted $\bar{\nu}_e$ event rate to the expected $\bar{\nu}_e$ in the case of no oscillations as a function of L_0/E in the KamLAND experiment. L_0 is the effective baseline calculated from the flux-weighted average distance of all the reactors. Reprinted figure with permission from S. Abe et al. (KamLAND Collaboration), Phys. Rev. Lett., vol. 100, p. 221803, 2008. Copyright (2008) by the American Physical Society [55]

The Irvine-Michigan-Brookhaven (IMB) experiment in the United States and the Kamiokande experiment in Japan both observed a smaller number of muon-like neutrino events than expected [56, 57].

In 1998 the Super-Kamiokande detector—a 50 kiloton water Cherenkov detector deep under a mountain—measured a substantially different ratio of muon-like to electron-like events than predicted by simulations. The use of Cherenkov light allowed Super-Kamiokande to determine the direction of the incoming neutrino. A neutrino arriving from below the detector had been produced in the atmosphere on the other side of the Earth and hence had travelled thousands of kilometres. In contrast, a neutrino arriving from above only travelled a few kilometres. Super-Kamiokande measured a substantial depletion in the muon neutrino event rate arriving from below the detector, whilst the number of electron neutrino events was relatively constant from all directions. Figure 2.5, taken from the relevant paper, clearly shows this effect. This result indicated the existence of $\nu_\mu \rightarrow \nu_\tau$ oscillations and provided the first measurements of the atmospheric mixing parameters: $\sin^2 2\theta_{23} > 0.82$ and 5×10^{-4} eV^2 $< \Delta m^2_{32} < 6 \times 10^{-3}$ eV^2 at 90% confidence level [31].

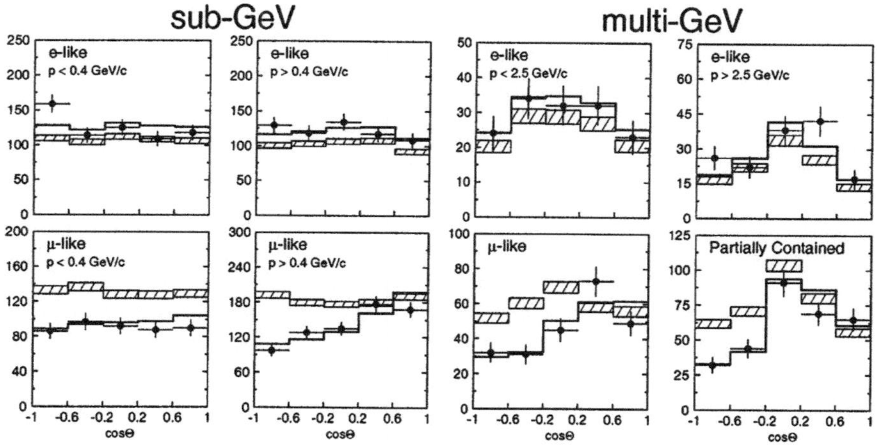

Fig. 2.5 Atmospheric neutrino results from Super-Kamiokande. The sub-GeV and multi-GeV μ-like and e-like events are plotted as a function of the zenith angle. The data points are plotted alongside the Monte Carlo prediction assuming no neutrino oscillations in the hatched boxes. A line is fitted to the data that assumes $\nu_\mu \rightarrow \nu_\tau$ oscillations occur. Reprinted figure with permission from Y. Fukuda, et al. (Super-Kamiokande Collaboration), Phys. Rev. Lett., vol. 81, pp. 1562—1567, 1998. Copyright (1998) by the American Physical Society [31]

2.4.3 Reactor Neutrinos

As mentioned in the discussion on the KamLAND experiment, electron antineutrinos from nuclear reactors provide a stable and relatively abundant source of electron antineutrinos to study oscillations. The energy scale of reactor neutrinos is too low to allow CC muon or tau neutrino interactions, hence reactor neutrino experiments measure the disappearance of the electron antineutrinos. Electron antineutrinos at MeV energies are typically measured through inverse beta decay ($\bar{\nu}_e + p \rightarrow n + e^+$), where the positron annihilates to produce a photon, followed by a second photon emission from the delayed capture of the neutron. It was reactor neutrino experiments that provided the first measurements of a non-zero value for θ_{13}.

The Daya Bay experiment used six liquid scintillator detectors to measure $\bar{\nu}_e$ disappearance from six nuclear reactors at an average distance of 1 km, which was significantly shorter than KamLAND baseline. Detectors close to the nuclear reactors measured the initial antineutrino flux before oscillations and the detectors positioned at a greater distance measured the remaining antineutrino flux after oscillations. Daya Bay measured the mixing angle to be $\sin^2 2\theta_{13} = 0.092 \pm 0.016$ and the other reactor neutrino experiments operating at the same time as Daya Bay – RENO and Double Chooz – also made consistent measurements of θ_{13} [58–60].

A non-zero value for all the mixing angles, including θ_{13}, is a necessary pre-condition for CP violation in neutrino oscillations. Therefore, the measurement of

θ_{13} opens the way for experiments searching for CP violation in the neutrino sector and currently long-baseline accelerator neutrino beam experiments are the leading method for making such a measurement.

2.4.4 Accelerator Neutrinos

A neutrino beam produced by a particle accelerator provides an experiment with an intense source of high-energy neutrinos at a range of neutrino energies appropriately chosen for the oscillation parameters of interest. As discussed in Sect. 2.1, the first artificial neutrino beam was produced at Brookhaven Laboratory by directing a proton beam on to a beryllium target [21]. This method was greatly enhanced by the development of the magnetic focusing horn by Simon van der Meer [61], which employs magnetic fields to direct the charged hadrons into a collimated beam that subsequently decays into neutrinos. Since most of the hadrons are pions that decay as $\pi \rightarrow \mu\nu_{\mu}$, the resulting focused neutrino beam is predominantly composed of muon neutrinos. The direction of the magnetic field can be flipped in order to focus positive charge particles and de-focus negative charge particles and vice versa. Therefore, the focusing of positive charge particles results in a predominantly muon neutrino beam and the focusing of negative charge particles results in a predominantly muon antineutrino beam. Further details on accelerator neutrino beams will be provided in Chap. 3.

The first long baseline accelerator neutrino oscillation experiments were K2K and MINOS, in which the neutrinos travelled 250 km and 735 km respectively from the neutrino beam source to the detectors [62, 63]. Both experiments made consistent measurements of Δm^2_{32} and θ_{23} by observing the disappearance of muon neutrinos. The disappearance of muon neutrinos in MINOS is shown clearly in Fig. 2.6. Searches for electron neutrino appearance were also conducted, which provided sensitivity to θ_{13} [64, 65]. K2K and MINOS used a near and far detector, where the "far detector" is positioned far from the neutrino beam source and measures the oscillated neutrinos and the "near detector" is close to the neutrino beam source and measures neutrinos before oscillation with high statistics in order to control systematic uncertainties. This experimental configuration has been adopted by all subsequent long baseline neutrino oscillation experiments.

The T2K and NOνA experiments are the successor experiments to K2K and MINOS that are currently operational. T2K produces an intense neutrino beam from 30 GeV protons, which is directed 295 km across Japan to the Super Kamiokande water Cherenkov detector [67]. NOνA on the other hand uses a higher energy neutrino beam and longer baseline in conjunction with a liquid scintillator detector technology [68, 69].

In addition to muon neutrino disappearance and the Δm^2_{32} and θ_{23} parameters, T2K and NOνA were designed to provide greater sensitivity to $\nu_{\mu} \rightarrow \nu_e$ and $\bar{\nu}_{\mu} \rightarrow \bar{\nu}_e$ appearance. In 2014 T2K presented its first measurement of $\nu_{\mu} \rightarrow \nu_e$ appearance in a neutrino beam, providing independent confirmation for a non-zero value of θ_{13} [70].

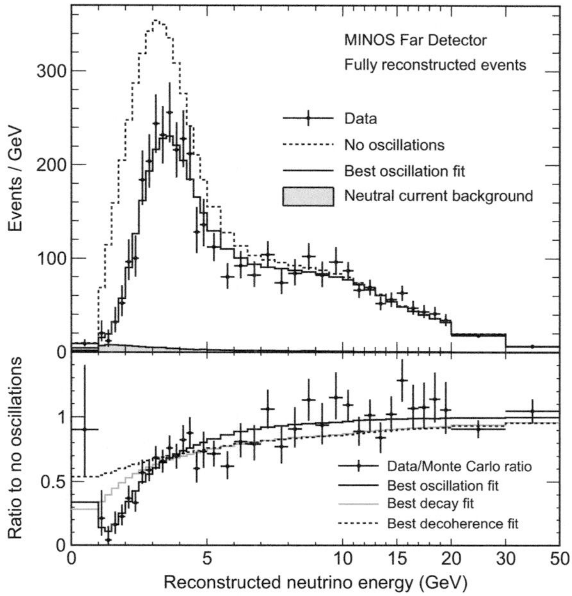

Fig. 2.6 Fully reconstructed muon neutrino event rate at the MINOS far detector and the expected event rate in the case of no oscillations. The bottom plot is the ratio of the background subtracted data to the no oscillations hypothesis prediction. Reprinted figure with permission from P. Adamson, et al. (MINOS Collaboration), Phys. Rev. Lett., vol. 106, p. 181801, 2011. Copyright (2011) by the American Physical Society [66]

Sensitivity to ν_e and $\bar{\nu}_e$ appearance also allows accelerator neutrino experiments to make measurements of δ_{CP}, which would establish whether CP symmetry is violated in neutrino oscillations. CP violation manifests itself in the observation of different oscillation probabilities for $\nu_\mu \rightarrow \nu_e$ and $\bar{\nu}_\mu \rightarrow \bar{\nu}_e$ transitions and the first measurement of δ_{CP} was presented by T2K in 2020 [71]. As discussed in Sect. 2.3.3, different oscillation probabilities for neutrinos and antineutrinos can also be caused by the MSW effect, since the neutrinos travel through the Earth's crust in long baseline neutrino oscillation experiments. However, due to the energy range of the neutrino beam and relatively short baseline, this is not a large effect for T2K.

Establishing whether CP symmetry is indeed violated in neutrino oscillations and making precise measurements of all the oscillation parameters requires a future generation of accelerator neutrino experiments. This next-generation will be the Deep Underground Neutrino Experiment (DUNE) and Hyper Kamiokande and this thesis will focus on DUNE [72, 73]. DUNE will measure muon neutrino disappearance and electron neutrino appearance over a very long baseline (1285 km) and in a relatively high-energy beam. A full description of DUNE is provided in Chap. 3.

2.4.5 *Overview of Oscillation Parameter Measurements*

A combination of solar, atmospheric, reactor and accelerator neutrino oscillation experiments have provided a comprehensive set of measurements for most parameters in the PMNS matrix and the neutrino mass-squared differences. A joint fit of the solar neutrino measurements from SNO and reactor neutrino measurements from KamLAND have confirmed that $\Delta m^2_{21} > 0$, meaning the mass of the ν_2 state is greater than the mass of ν_1.

The sign of the Δm^2_{32} is currently poorly constrained. Future long baseline neutrino oscillation experiments, such as DUNE, aim to determine whether the neutrino masses have *normal ordering*, $\Delta m^2_{32} > 0$, or *inverted ordering*, $\Delta m^2_{32} < 0$. These two scenarios for the neutrino mass splitting are illustrated in Fig. 2.7. The sign of Δm^2_{32} will be determined by measuring oscillation probabilities in matter. If the neutrino mass states have normal ordering, the neutrino oscillation probability will be enhanced relative to the antineutrinos. The opposite is true for inverted ordering. The long baseline and high neutrino energies of DUNE make it an ideal experiment to measure the neutrino mass ordering. Figure 2.8 shows the oscillation probabili-

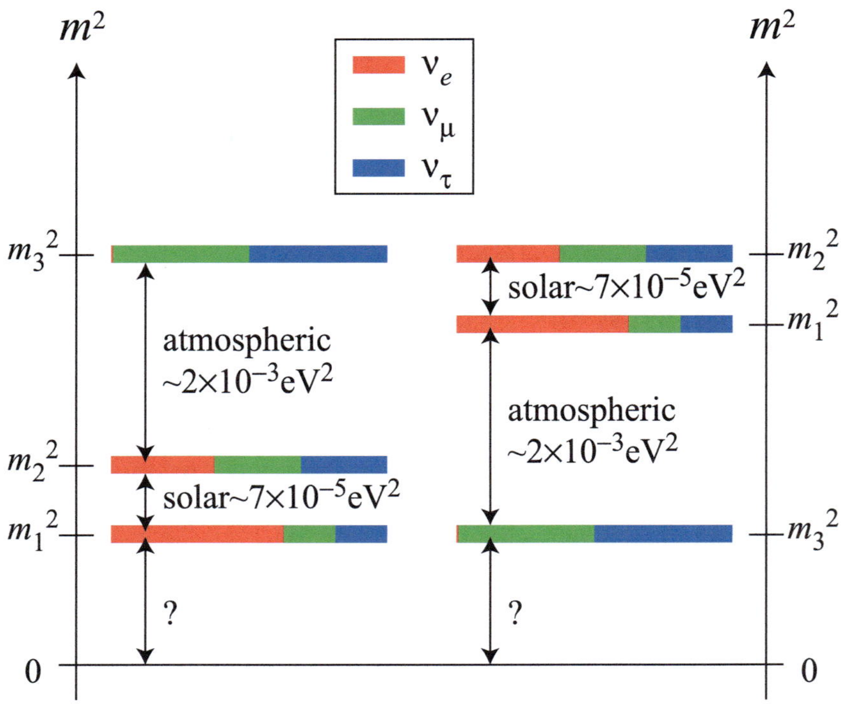

Fig. 2.7 Diagram illustrating the two scenarios for the ordering of the neutrino masses. Either ν_3 is the heaviest state (normal ordering) or the lightest state (inverted ordering). [74]

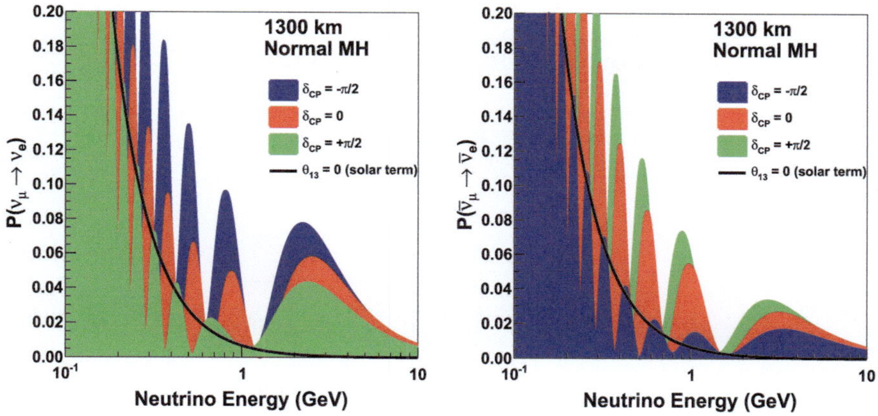

Fig. 2.8 Oscillation probabilities for neutrinos and antineutrinos in the Earth's crust, assuming normal mass ordering and a 1300 km baseline. Probabilities for different values of δ_{CP} are shown. The oscillation probability for neutrinos at the first oscillation maximum (at approximately 2.5 GeV) is enhanced relative to the antineutrino case. Figure taken from Reference [75]

ties for neutrinos and antineutrinos in matter, assuming normal mass ordering and a 1300 km baseline. At the first oscillation maximum a clear enhancement can be seen in the neutrino oscillation probability relative to the antineutrinos and this is due to the MSW effect in the Earth's crust.

A summary of the current measured parameters of the three-flavour oscillation model is provided by the Particle Data Group (PDG) [76]. PDG draws on global fit results from NuFIT [77]. The best fit value of each parameter, its uncertainty and the main type of neutrino experiment (as categorised in Sect. 2.4) used to measure that parameter are summarised in Table 2.1. In addition, contours that show the current status of measurements of Δm^2_{32} and $\sin^2\theta_{23}$ are displayed in Fig. 2.9. Accurate measurements of these parameters in a long baseline oscillation experiment is a particular focus of this thesis.

Table 2.1 Best fit values and uncertainties of the three-flavour oscillation model parameters for normal ordering ($\Delta m^2_{32} > 0$) and inverted ordering ($\Delta m^2_{32} < 0$). Results are from References [76] and [77]

	Experiment type	Normal ordering	Inverted ordering
$\sin^2\theta_{12}$	Solar	$0.304^{+0.013}_{-0.012}$	$0.304^{+0.013}_{-0.012}$
$\sin^2\theta_{23}$	Accel.	$0.570^{+0.018}_{-0.024}$	$0.575^{+0.017}_{-0.021}$
$\sin^2\theta_{13}$	Reactor	$0.02221^{+0.00068}_{-0.00062}$	$0.02240^{+0.00062}_{-0.00062}$
δ_{CP}	Accel.	195^{+51}_{-25} °	286^{+27}_{-32} °
Δm^2_{21}	Reactor/Solar	$7.42^{+0.21}_{-0.20} \times 10^{-5}\text{eV}^2$	$7.42^{+0.21}_{-0.20} \times 10^{-5}\text{eV}^2$
Δm^2_{32}	Accel./Reactor	$+2.514^{+0.028}_{-0.027} \times 10^{-3}\text{eV}^2$	$-2.497^{+0.028}_{-0.028} \times 10^{-3}\text{eV}^2$

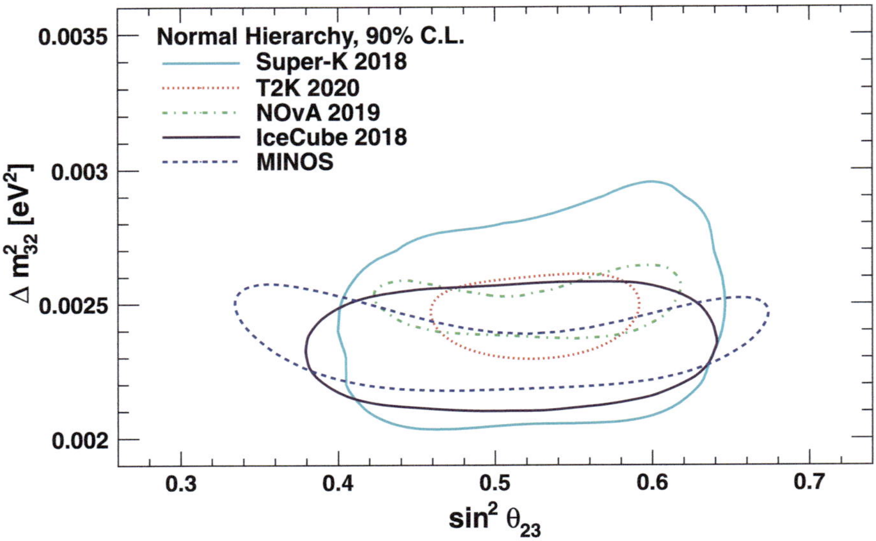

Fig. 2.9 Current status of measurements of Δm^2_{32} and $\sin^2 \theta_{23}$. The precision of the measurement of these parameters is currently led by the long baseline experiments. Reprinted from Progress in Particle and Nuclear Physics, vol. 124, M. S. Athar et al., "Status and perspectives of neutrino physics," pp. 103947, Copyright (2022), with permission from Elsevier [78]

2.5 Neutrinos as a Window to New Physics

The original formulation of the Glashow-Weinberg-Salam Standard Model assumed the neutrinos to be massless particles. This was largely a legacy of the two-component neutrino theory developed by Landau [79], Lee and Yang [80] and Salam [81] in the 1950s to explain the non-conservation of parity in weak interactions by requiring the neutrino to be massless. Although parity violation was eventually incorporated into the Standard Model through the V-A theory (see Sect. 2.2.1), the massless neutrino assumption persisted. The discovery of neutrino oscillations overturned that assumption, requiring the Standard Model to accommodate neutrino flavour mixing as a consequence of the neutrinos having mass. Massive neutrinos and the mixing parameters of the PMNS matrix have the potential to reveal new physics beyond the Standard Model and improve our understanding of the origin of the matter-antimatter asymmetry in the early universe.

A Dirac fermion is a spin-1/2 particle that is distinct from its corresponding antiparticle. This applies to the charged leptons and quarks of the Standard Model. Whilst the neutrinos have a non-zero mass, it is many orders of magnitude smaller than the other fermions. If neutrinos were Dirac particles it would require the neutrinos to have very small Yukawa couplings to the Higgs field. An alternative explanation for the small neutrino masses is the so-called seesaw mechanism [82, 83]. This mechanism relies on the neutrinos being Majorana rather than Dirac particles, where an additional Majorana mass term is included in the Standard Model Lagrangian. A Majorana particle is identical to its antiparticle, which implies that lepton number is not conserved as a fundamental symmetry. In the seesaw mechanism the right-handed neutrino and left-handed antineutrino, which do not couple to the electroweak interaction, have Majorana masses far larger than the Dirac masses, resulting in small Majorana masses for the active neutrinos. In this hypothesis the Dirac masses of the neutrinos are a similar order of magnitude to the other fermions, thereby providing a natural explanation for the smallness of the observable neutrino masses.

Experimental tests of the Dirac or Majorana nature of the neutrino have focused on searches for neutrinoless double β-decay [84, 85]. This lepton number violating process will only occur if neutrinos are Majorana particles. The mass ordering measurement carried out by DUNE has significant implications for neutrinoless double β-decay searches. If DUNE were to measure the neutrino mass ordering to be inverted, then it would be possible to confirm or refute the existence of neutrinoless double β-decay within the next generation of experiments and thereby determine whether the neutrinos are Majorana or Dirac particles [86].

The observed matter-antimatter asymmetry refers to an asymmetry in the number of baryons and anti-baryons in the universe. Understanding the source of the baryon asymmetry is one of the greatest challenges in particle physics and cosmology, as discussed in Chap. 1. A promising solution is a leptogenesis model in which a lepton asymmetry, brought about through CP-violating decays of right-handed neutrinos, is converted into a baryon asymmetry [87]. The extent to which the CP-violating phase δ_{CP} measured by neutrino oscillation experiments is connected to the baryon asymmetry is dependent on the choice of leptogenesis model. Despite this, evidence of CP violation in neutrino oscillations in conjunction with neutrinos being Majorana particles would undoubtedly favour leptogensis as a leading theory for the origin of the baryon asymmetry. It is therefore vital that CP violation in the neutrino sector is fully explored.

2.6 Modelling Neutrino Interactions

Accurately modelling neutrino-nucleus interactions is of great importance for the next generation of long baseline neutrino oscillation experiments. The neutrino interaction model (sometimes referred to as the cross section model) relates the initial neutrino energy to the final state particles in an event. Currently, the neutrino interaction model is expected to be one of the most challenging sources of systematic uncertainty in DUNE and this is unlikely to change on the time scale of the experiment [88]. The importance of accurately modelling neutrino-nucleus interactions arises from the fact that neither the energy of the incoming neutrino nor the kinematics of the interacting nucleon within the nucleus are known experimentally. The neutrino interaction model, implemented in Monte Carlo (MC) simulations of neutrino interactions, is therefore relied upon to understand the mapping of the true kinematics of the incoming neutrino to the observable final state in the detector. Any

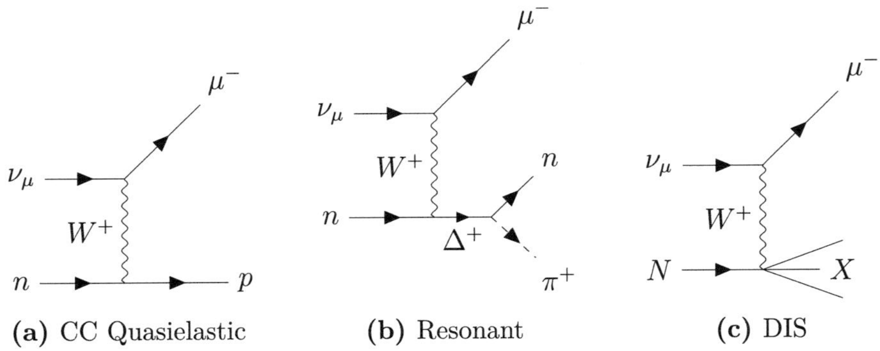

Fig. 2.10 Feynman diagrams for the three main neutrino interaction types relevant to long baseline neutrino oscillation experiments

inaccuracies or biases in this mapping can distort the distribution of reconstructed neutrino energies, leading to an inaccurate measurement of the oscillation probability [89–92].

A neutrino interacting with a large nucleus at the energy scales relevant to long baseline oscillation experiments is, in general, scattering on a single nucleon within the nuclear medium. Hence, the first task of the neutrino interaction model is to calculate the cross section for a neutrino-nucleon interaction. This is challenging because the interactions of interest to oscillation experiments occur at energy scales spanning several orders of magnitude, from 10 s of MeV to 1000 s of MeV [93]. Consequently, there are contributions to the total neutrino cross section from several types of neutrino-nucleon interactions. The three most common CC neutrino interaction types in this energy range are illustrated in Fig. 2.10.

The simplest CC interaction is a neutrino scattering off a single nucleon, which is then ejected from the nucleus. Due to the non-conservation of lepton masses, this process is called "charged current quasielastic" (CCQE) scattering. At slightly higher energies it is possible for the neutrino CC interaction to excite the nucleon into a resonant state, which then decays to a variety of hadronic states, typically including one nucleon and one pion. Such processes are called "resonant" scattering. Finally, deep inelastic scattering (DIS) can occur, where the neutrino interacts with the individual quarks, breaking apart the nucleon and creating a hadronic shower. DIS dominates at the highest neutrino energies. Figure 2.11 illustrates how, for a given neutrino energy, the total neutrino cross section may have contributions from several types of neutrino interaction. Consequently, the neutrino interaction model must be capable of accurately predicting the neutrino-nucleon cross section for several different types of interaction as well as the relative contribution of each process to the total cross section across a broad spectrum of energies.

The neutrino interaction model must also account for neutrino-nucleon interactions occurring within the nuclear medium. Complex nuclear effects can substantially alter the cross section of the neutrino interaction and the kinematics and multiplicity

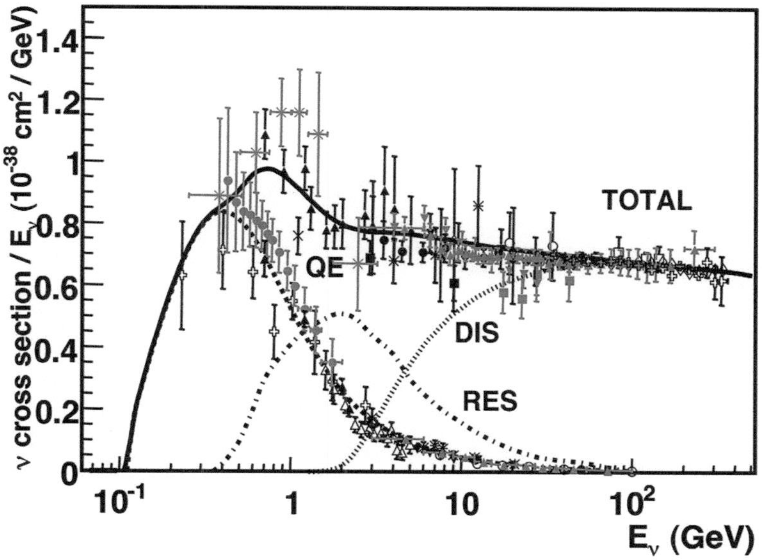

Fig. 2.11 Total muon neutrino CC cross section per nucleon. Contributions from different interaction types are shown. CCQE events dominate at low energies and DIS dominates at high energies. Reprinted figure with permission from J. A. Formaggio and G. P. Zeller, Rev. Mod. Phys., vol. 84, pp. 1307—1341, 2012. Copyright (2012) by the American Physical Society [93]

of the final state particles observable in the detector. Furthermore, no unified nuclear model exists that can account for all the potential sources of systematic uncertainty. Instead, a plethora of models are combined to simulate the neutrino scattering process. The initial state of the nucleus is typically described by the relativistic Fermi-gas model [94], However there are more sophisticated approaches such as spectral functions [95] and random phase approximation (RPA) calculations [96]. Experimentally, a CCQE event is identified by a single lepton and nucleon in the detector, although the nucleon is often not well reconstructed. The sample of CCQE events can have contributions from so-called 2 particle-2 hole (2p-2h) events, where the neutrino interacts with a correlated pair of nucleons, which significantly alters the cross section from the single nucleon case [97–100]. Cross sections for resonance production processes, such as the one shown in Fig. 2.10b , are commonly calculated using the Rein and Sehgal model [101].

Efforts to reconstruct neutrino energies in scattering events across such a broad spectrum of energies are complicated by the presence of neutral particles, misidentified particles and final state interactions (FSI). Once a neutrino interacts within the nucleus the final state particles must first traverse the nuclear medium before they can be observed in the detector. Typically, the lepton escapes the nucleus. The hadrons however can be be scattered, absorbed and re-emitted by the nuclear medium, which can alter the apparent neutrino energy measured in the detector. Intra-nuclear cascade models are a popular choice for modelling FSI in neutrino-nucleus interactions [102],

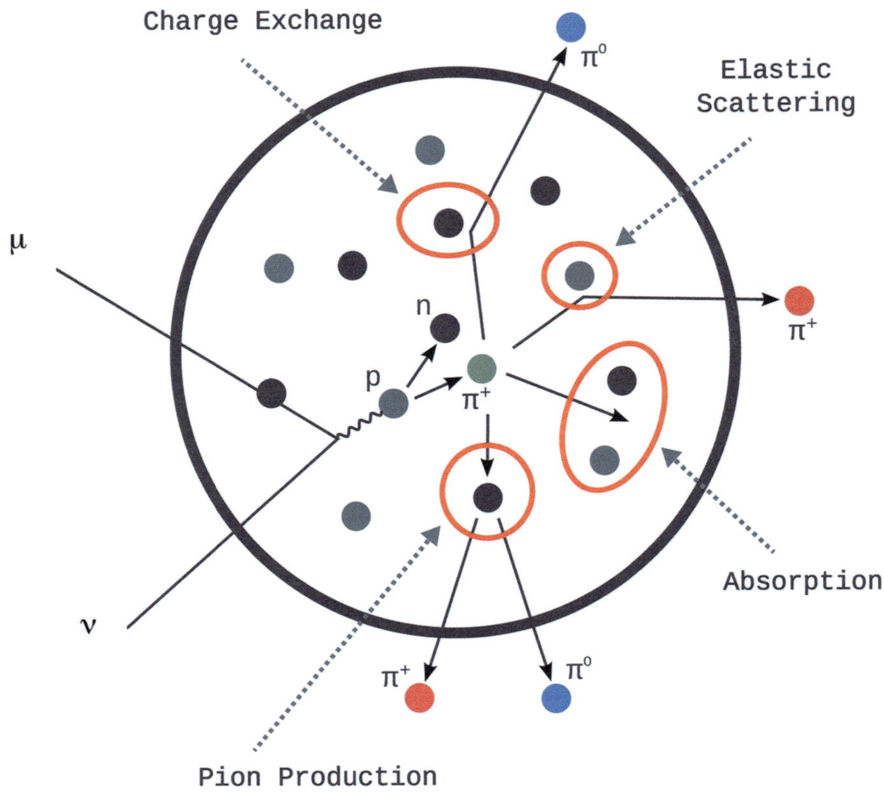

Fig. 2.12 Illustration of how FSI effects can alter the kinematics and multiplicities of observable hadrons in the detector. Pions are often absorbed by the nuclear medium, potentially biasing the neutrino energy estimation. Figure produced by Tomasz Golan. Reprinted from Progress in Particle and Nuclear Physics, vol. 100,L. Alvarez-Ruso et al. (Neutrino Scattering Theory Experiment Collaboration), "NuSTEC White Paper: Status and challenges of neutrino–nucleus scattering," pp. 1—68, Copyright (2018), with permission from Elsevier [88]

although other models are available. An example of how FSI can affect the observable particles in a neutrino event is illustrated in Fig. 2.12. Furthermore, after exiting the nucleus, final state neutral hadrons, such as neutrons, may go undetected and charged hadrons may be misidentified, leading to a biased neutrino energy estimation.

There is a wide program of experiments that are attempting to better measure neutrino cross sections. These measurements aim to provide more accurate MC simulations that are used to extract the values of the oscillation parameters. Such past, present and future experiments include MiniBooNE [103], MINERνA [104], MicroBooNE [105] and SBND [106]. In addition, there are techniques available, such as measuring the transverse kinematic imbalance (TKI) of a neutrino event [107, 108], that aim to directly measure the nuclear effects in a neutrino-nucleus interaction.

The long baseline oscillation analysis presented in this thesis employs a different approach, where the neutrino interaction model is no longer relied upon to extract the oscillation parameters from the observable quantities in the detector. Instead a data-driven approach is presented that attempts to build predictions of the oscillated event rates directly from data.

References

1. F. Close, Neutrino (Oxford University Press, Oxford, UK, 2010)
2. F. Reines, The early days of experimental neutrino physics. Science **203**(4375), 11–16 (1979). https://www.science.org/doi/pdf/10.1126/science.203.4375.11
3. D.J. Griffiths, Introduction to elementary particles; 2nd rev. version (Physics textbook) (Wiley, New York, NY, 2008)
4. B.R. Martin, G. Shaw, *Particle Physics (Physics textbook)*, 4th edn. (Wiley, New York, NY, 2017)
5. M. Thomson, *Modern Particle Physics* (Cambridge University Press, Cambridge, UK, 2013)
6. B. Kayser, "Neutrino physics," eConf, vol. C040802, J. Hewett, J. Jaros, T. Kamae, and C. Prescott, Eds., p. L004, 2004. arXiv: hep-ph/0506165
7. B. Kayser, "Neutrino Oscillation Physics," in 2011 European School of High-Energy Physics, 2014, pp. 107–117. arXiv: 1206.4325 [hep-ph]
8. R.N. Cahn, G. Goldhaber, *The Experimental Foundations of Particle Physics*, 2nd edn. (Cambridge University Press, Cambridge, UK, 2009)
9. J. Chadwick, The intensity distribution in the magnetic spectrum of beta particles from radium (B + C). Verh. Phys. Gesell. **16**, 383–391 (1914)
10. W. Pauli, Letter of the 4th december 1930
11. J. Chadwick, Possible Existence of a Neutron. Nature **129**, 312 (1932)
12. E. Fermi, Versuch einer Theorie der β-Strahlen. I. Z. Phys. **88**, 161–177 (1934)
13. H. Bethe, R. Peierls, The "Neutrino". Nature **133**, 532 (1934)
14. F. Reines and C. L. Cowan, "Detection of the free neutrino," Phys. Rev., vol. 92, pp. 830–831, 3 Nov. 1953
15. C.L. Cowan et al., Detection of the free neutrino: A confirmation. Science **124**(3212), 103–104 (1956)
16. M. Goldhaber, L. Grodzins, and A. W. Sunyar, "Helicity of neutrinos," Phys. Rev., vol. 109, pp. 1015–1017, 3 Feb. 1958
17. E. C. G. Sudarshan and R. E. Marshak, "Chirality invariance and the universal fermi interaction," Phys. Rev., vol. 109, pp. 1860–1862, 5 Mar. 1958
18. R. P. Feynman and M. Gell-Mann, "Theory of the fermi interaction," Phys. Rev., vol. 109, pp. 193–198, 1 Jan. 1958
19. B. Pontecorvo, Electron and Muon Neutrinos. Zh. Eksp. Teor. Fiz. **37**, 1751–1757 (1959)
20. M. Schwartz, "Feasibility of using high-energy neutrinos to study the weak interactions," Phys. Rev. Lett., vol. 4, pp. 306–307, 6 Mar. 1960
21. G. Danby et al., "Observation of high-energy neutrino reactions and the existence of two kinds of neutrinos," Phys. Rev. Lett., vol. 9, pp. 36–44, 1 Jul. 1962
22. M. L. Perl et al., "Evidence for anomalous lepton production in e^+-e^- annihilation," Phys. Rev. Lett., vol. 35, pp. 1489–1492, 22 Dec. 1975
23. K. Kodama et al., Observation of tau neutrino interactions. Phys. Lett. B **504**(3), 218–224 (2001)
24. R. Davis, D. S. Harmer, and K. C. Hoffman, "Search for neutrinos from the sun," Phys. Rev. Lett., vol. 20, pp. 1205–1209, 21 May 1968

25. J. N. Bahcall, "Solar neutrinos. i. theoretical," Phys. Rev. Lett., vol. 12, pp. 300–302, 11 Mar. 1964
26. P. Anselmann et al., Gallex results from the first 30 solar neutrino runs. Phys. Lett. B **327**(3), 377–385 (1994)
27. V. Gavrin et al., Sage: The soviet-american gallium solar neutrino experiment. Nucl. Phys. B Proc.Suppl. **28**(1), 75–81 (1992)
28. B. Pontecorvo, "Mesonium and antimesonium," Zh. Eksp. Teor. Fiz., vol. 33, 1957
29. B. Pontecorvo, Inverse beta processes and nonconservation of lepton charge. Zh. Eksp. Teor. Fiz. **34**, 247 (1957)
30. B. Pontecorvo, Neutrino Experiments and the Problem of Conservation of Leptonic Charge. Zh. Eksp. Teor. Fiz. **53**, 1717–1725 (1967)
31. Super-Kamiokande Collaboration, Y. Fukuda, et al., "Evidence for oscillation of atmospheric neutrinos," Phys. Rev. Lett., vol. 81, pp. 1562–1567, 8 Aug. 1998
32. SNO Collaboration, Q. R. Ahmad, et al., "Direct evidence for neutrino flavor transformation from neutral-current interactions in the sudbury neutrino observatory," Phys. Rev. Lett., vol. 89, p. 011 301, 1 Jun. 2002
33. KamLAND Collaboration, K. Eguchi, et al., "First results from kamland: Evidence for reactor antineutrino disappearance," Phys. Rev. Lett., vol. 90, p. 021 802, 2 Jan. 2003
34. C. S. Wu et al., "Experimental test of parity conservation in beta decay," Phys. Rev., vol. 105, pp. 1413–1415, 4 Feb. 1957
35. S.L. Glashow, Partial-symmetries of weak interactions. Nucl. Phys. **22**(4), 579–588 (1961)
36. A. Salam, J. Ward, Electromagnetic and weak interactions. Phys. Lett. **13**(2), 168–171 (1964)
37. S. Weinberg, "A model of leptons," Phys. Rev. Lett., vol. 19, pp. 1264–1266, 21 Nov. 1967
38. F. Englert and R. Brout, "Broken symmetry and the mass of gauge vector mesons," Phys. Rev. Lett., vol. 13, pp. 321–323, 9 Aug. 1964
39. P. W. Higgs, "Broken symmetries and the masses of gauge bosons," Phys. Rev. Lett., vol. 13, pp. 508–509, 16 Oct. 1964
40. G. S. Guralnik, C. R. Hagen, and T. W. B. Kibble, "Global conservation laws and massless particles," Phys. Rev. Lett., vol. 13, pp. 585–587, 20 Nov. 1964
41. F. Hasert et al., Observation of neutrino-like interactions without muon or electron in the gargamelle neutrino experiment. Phys. Lett. B **46**(1), 138–140 (1973)
42. G. Arnison et al., Experimental observation of isolated large transverse energy electrons with associated missing energy at s=540 gev. Phys. Lett. B **122**(1), 103–116 (1983)
43. M. Banner et al., Observation of single isolated electrons of high transverse momentum in events with missing transverse energy at the cern pp collider. Phys. Lett. B **122**(5), 476–485 (1983)
44. "Precision electroweak measurements on the z resonance," Physics Reports, vol. 427, no. 5, pp. 257–454, 2006
45. N. Cabibbo, "Unitary symmetry and leptonic decays," Phys. Rev. Lett., vol. 10, pp. 531–533, 12 Jun. 1963
46. M. Kobayashi and T. Maskawa, "CP-Violation in the Renormalizable Theory of Weak Interaction," Progress of Theoretical Physics, vol. 49, no. 2, pp. 652–657, Feb. 1973. eprint: https://academic.oup.com/ptp/article-pdf/49/2/652/5257692/49-2-652.pdf
47. Z. Maki, M. Nakagawa, S. Sakata, Remarks on the unified model of elementary particles. Prog. Theor. Phys. **28**, 870–880 (1962)
48. H. Nunokawa, S. J. Parke, and J. W. F. Valle, "CP Violation and Neutrino Oscillations," Prog. Part. Nucl. Phys., vol. 60, pp. 338–402, 2008. arXiv: 0710.0554 [hep-ph]
49. L. Wolfenstein, "Neutrino oscillations in matter," Phys. Rev. D, vol. 17, pp. 2369–2374, 9 May 1978
50. S. P. Mikheev and A. Y. Smirnov, "Resonance enhancement of oscillations in matter and solar neutrino spectroscopy," Sov. J. Nucl. Phys. (Engl. Transl.); (United States), vol. 42:6, Dec. 1985
51. E. K. Akhmedov, "Neutrino physics," in ICTP Summer School in Particle Physics, Jun. 1999, pp. 103–164. arXiv: hep-ph/0001264

52. Super-Kamiokande Collaboration, S. Fukuda, et al., "Solar ^8B and hep Neutrino Measurements from 1258 Days of Super-Kamiokande Data," Phys. Rev. Lett., vol. 86, pp. 5651–5655, 25 Jun. 2001
53. SNO Collaboration, Q. R. Ahmad, et al., "Measurement of the Rate of $\nu_e + d \rightarrow p + p + e^-$ Interactions Produced by ^8B Solar Neutrinos at the Sudbury Neutrino Observatory," Phys. Rev. Lett., vol. 87, p. 071 301, 7 Jul. 2001
54. SNO Collaboration, B. Aharmim, et al., "Electron energy spectra, fluxes, and day-night asymmetries of ^{8}B solar neutrinos from measurements with NaCl dissolved in the heavy-water detector at the Sudbury Neutrino Observatory," Phys. Rev. C, vol. 72, p. 055 502, 5 Nov. 2005
55. KamLAND Collaboration, S. Abe, et al., "Precision measurement of neutrino oscillation parameters with kamland," Phys. Rev. Lett., vol. 100, p. 221 803, 22 Jun. 2008
56. T. J. Haines et al., "Calculation of atmospheric neutrino-induced backgrounds in a nucleon-decay search," Phys. Rev. Lett., vol. 57, pp. 1986–1989, 16 Oct. 1986
57. K. Hirata et al., Experimental study of the atmospheric neutrino flux. Phys. Lett. B **205**(2), 416–420 (1988)
58. F. P. An et al., "Observation of electron-antineutrino disappearance at daya bay," Phys. Rev. Lett., vol. 108, p. 171 803, 17 Apr. 2012
59. RENO Collaboration, J. K. Ahn, et al., "Observation of reactor electron antineutrinos disappearance in the reno experiment," Phys. Rev. Lett., vol. 108, p. 191 802, 19 May 2012
60. Double Chooz Collaboration, Y. Abe, et al., "Indication of reactor $\overline{\nu}_e$ disappearance in the double chooz experiment," Phys. Rev. Lett., vol. 108, p. 131 801, 13 Mar. 2012
61. S. van der Meer, "A Directive Device for Charged Particles and Its use in an Enhanced Neutrino Beam," Tech. Rep. CERN-61-07, Feb. 1961
62. M. H. Ahn et al., "Indications of neutrino oscillation in a 250 km long-baseline experiment," Phys. Rev. Lett., vol. 90, p. 041 801, 4 Jan. 2003
63. MINOS Collaboration, D. G. Michael, et al., "Observation of muon neutrino disappearance with the minos detectors in the numi neutrino beam," Phys. Rev. Lett., vol. 97, p. 191 801, 19 Nov. 2006
64. K2K Collaboration, M. H. Ahn, et al., "Search for electron neutrino appearance in a 250 km long-baseline experiment," Phys. Rev. Lett., vol. 93, p. 051 801, 5 Jul. 2004
65. MINOS Collaboration, P. Adamson, et al., "Electron Neutrino and Antineutrino Appearance in the Full MINOS Data Sample," Phys. Rev. Lett., vol. 110, no. 17, p. 171 801, 2013. arXiv: 1301.4581 [hep-ex]
66. MINOS Collaboration, P. Adamson, et al., "Measurement of the neutrino mass splitting and flavor mixing by minos," Phys. Rev. Lett., vol. 106, p. 181 801, 18 May 2011
67. T2K Collaboration, K. Abe, et al., "The T2K Experiment," Nucl. Instrum. Meth. A, vol. 659, pp. 106–135, 2011. arXiv: 1106.1238 [physics.ins-det]
68. D. S. Ayres et al., "The NOvA Technical Design Report," Oct. 2007
69. M. A. Acero et al., "First Measurement of Neutrino Oscillation Parameters using Neutrinos and Antineutrinos by NOvA," Phys. Rev. Lett., vol. 123, no. 15, p. 151 803, 2019. arXiv: 1906.04907 [hep-ex]
70. T2K Collaboration, K. Abe, et al., "Observation of electron neutrino appearance in a muon neutrino beam," Phys. Rev. Lett., vol. 112, p. 061 802, 6 Feb. 2014
71. T2K Collaboration, K. Abe, et al., "Constraint on the matter–antimatter symmetry-violating phase in neutrino oscillations," Nature, vol. 580, no. 7803, pp. 339–344, 2020, [Erratum: Nature 583, E16 (2020)]. arXiv: 1910.03887 [hep-ex]
72. DUNE Collaboration, B. Abi, et al., "Long-baseline neutrino oscillation physics potential of the DUNE experiment," Eur. Phys. J. C, vol. 80, no. 10, p. 978, 2020. arXiv: 2006.16043 [hep-ex]
73. K. Abe et al., "Physics potential of a long-baseline neutrino oscillation experiment using a J-PARC neutrino beam and Hyper-Kamiokande," PTEP, vol. 2015, p. 053C02, 2015. arXiv: 1502.05199 [hep-ex]
74. S. F. King and C. Luhn, "Neutrino Mass and Mixing with Discrete Symmetry," Rept. Prog. Phys., vol. 76, p. 056 201, 2013. arXiv: 1301.1340 [hep-ph]

75. DUNE Collaboration, B. Abi, et al., "Deep Underground Neutrino Experiment (DUNE), Far Detector Technical Design Report, Volume II: DUNE Physics," Feb. 2020. arXiv: 2002.03005 [hep-ex]
76. Particle Data Group, P. A. Zyla, et al., "Review of Particle Physics," Progress of Theoretical and Experimental Physics, vol. 2020, no. 8, Aug. 2020, 083C01. eprint: https://academic.oup.com/ptep/article-pdf/2020/8/083C01/34673722/ptaa104.pdf
77. I. Esteban et al., "The fate of hints: updated global analysis of three-flavor neutrino oscillations," JHEP, vol. 09, p. 178, 2020. arXiv: 2007.14792 [hep-ph]
78. M. S. Athar et al., "Status and perspectives of neutrino physics," Progress in Particle and Nuclear Physics, vol. 124, p. 103 947, 2022
79. L. Landau, On the conservation laws for weak interactions. Nucl. Phys. **3**(1), 127–131 (1957)
80. T. D. Lee and C. N. Yang, "Parity nonconservation and a two-component theory of the neutrino," Phys. Rev., vol. 105, pp. 1671–1675, 5 Mar. 1957
81. A. Salam, "On parity conservation and neutrino mass," Il Nuovo Cimento, vol. 5, pp. 299–301, 1 Jan. 1957
82. P. Minkowski, $\mu \rightarrow e\gamma$ at a Rate of One Out of 10^9 Muon Decays? Phys. Lett. B **67**, 421–428 (1977)
83. O. Miranda and J. Valle, "Neutrino oscillations and the seesaw origin of neutrino mass," Nuclear Physics B, vol. 908, pp. 436–455, 2016, Neutrino Oscillations: Celebrating the Nobel Prize in Physics 2015
84. V. Albanese et al., "The SNO+ experiment," JINST, vol. 16, no. 08, P08059, 2021. arXiv: 2104.11687 [physics.ins-det]
85. S. Abe et al., "Search for the majorana nature of neutrinos in the inverted mass ordering region with kamland-zen," Phys. Rev. Lett., vol. 130, p. 051 801, 5 Jan. 2023
86. M. Agostini et al., "Toward the discovery of matter creation with neutrinoless $\beta\beta$ decay," Rev. Mod. Phys., vol. 95, no. 2, p. 025 002, 2023. arXiv: 2202.01787 [hep-ex]
87. M. Fukugita, T. Yanagida, Barygenesis without grand unification. Phys. Lett. B **174**(1), 45–47 (1986)
88. L. Alvarez-Ruso et al., Nustec white paper: Status and challenges of neutrino-nucleus scattering. Prog. Part. Nucl. Phys. **100**, 1–68 (2018)
89. P. Coloma and P. Huber, "Impact of nuclear effects on the extraction of neutrino oscillation parameters," Physical Review Letters, vol. 111, no. 22, Nov. 2013
90. P. Coloma, P. Huber, C.-M. Jen, and C. Mariani, "Neutrino-nucleus interaction models and their impact on oscillation analyses," Physical Review D, vol. 89, no. 7, Apr. 2014
91. U. Mosel, Neutrino interactions with nucleons and nuclei: Importance for long-baseline experiments. Annu. Rev. Nucl. Part. Sci. **66**(1), 171–195 (2016). (Oct.)
92. A. M. Ankowski, O. Benhar, C. Mariani, and E. Vagnoni, "Effect of the $2p2h$ cross-section uncertainties on an analysis of neutrino oscillations," Physical Review D, vol. 93, no. 11, Jun. 2016
93. J. A. Formaggio and G. P. Zeller, "From ev to eev: Neutrino cross sections across energy scales," Rev. Mod. Phys., vol. 84, pp. 1307–1341, 3 Sep. 2012
94. R. Smith, E. Moniz, Neutrino reactions on nuclear targets. Nucl. Phys. B **43**, 605–622 (1972)
95. H. Nakamura, R. Seki, Quasi-elastic neutrino-nucleus scattering and spectral function. Nucl. Phys. B Proc.Suppl. **112**(1), 197–202 (2002)
96. V. Pandey et al., "Low-energy excitations and quasielastic contribution to electron-nucleus and neutrino-nucleus scattering in the continuum random-phase approximation," Phys. Rev. C, vol. 92, no. 2, p. 024 606, 2015. arXiv: 1412.4624 [nucl-th]
97. M. Martini, "Two Particle-Two Hole Excitations in Charged Current Quasielastic Neutrino-Nucleus Interactions," J. Phys. Conf. Ser., vol. 408, A. Blondel, I. Efthymiopoulos, and G. Prior, Eds., p. 012 041, 2013. arXiv: 1110.5895 [hep-ph]
98. M. Martini, M. Ericson, G. Chanfray, and J. Marteau, "Unified approach for nucleon knock-out and coherent and incoherent pion production in neutrino interactions with nuclei," Phys. Rev. C, vol. 80, p. 065 501, 6 Dec. 2009

99. J. Amaro et al., Meson-exchange currents and quasielastic neutrino cross sections in the superscaling approximation model. Phys. Lett. B **696**(1), 151–155 (2011)
100. J. Nieves, I. R. Simo, and M. J. V. Vacas, "Inclusive charged-current neutrino-nucleus reactions," Phys. Rev. C, vol. 83, p. 045 501, 4 Apr. 2011
101. D. Rein, L.M. Sehgal, Neutrino-excitation of baryon resonances and single pion production. Ann. Phys. **133**(1), 79–153 (1981)
102. G. D. Harp, "Extension of the isobar model for intranuclear cascades to 1 gev," Phys. Rev. C, vol. 10, pp. 2387–2396, 6 Dec. 1974
103. MiniBooNE Collaboration, A. A. Aguilar-Arevalo, et al., "First measurement of the muon neutrino charged current quasielastic double differential cross section," Phys. Rev. D, vol. 81, p. 092 005, 9 May 2010
104. MINERvA Collaboration, J. Devan, et al., "Measurements of the inclusive neutrino and antineutrino charged current cross sections in MINERvA using the low-ν flux method," Phys. Rev. D, vol. 94, p. 112 007, 11 Dec. 2016
105. MicroBooNE Collaboration, P. Abratenko, et al., "First measurement of energy-dependent inclusive muon neutrino charged-current cross sections on argon with the microboone detector," Phys. Rev. Lett., vol. 128, p. 151 801, 15 Apr. 2022
106. G. Valdiviesso, "Status of the Short-Baseline Near Detector at Fermilab," PoS, vol. NuFact2021, p. 184, 2022
107. X.-G. Lu et al., "Reconstruction of energy spectra of neutrino beams independent of nuclear effects," Phys. Rev. D, vol. 92, p. 051 302, 5 Sep. 2015
108. X.-G. Lu et al., "Measurement of nuclear effects in neutrino interactions with minimal dependence on neutrino energy," Phys. Rev. C, vol. 94, p. 015 503, 1 Jul. 2016

Chapter 3
Deep Underground Neutrino Experiment

The Deep Underground Neutrino Experiment (DUNE) is a next-generation neutrino oscillation experiment designed to observe CP violation in the lepton sector, determine the neutrino mass ordering, and precisely measure the oscillation parameters δ_{CP}, θ_{13}, θ_{23} and the magnitude of Δm^2_{32}. This physics program will be a major contribution to the completion of the three-flavour neutrino mixing model. In addition, there is a wider physics program that includes the observation of supernovae neutrinos, proton decay detection and beyond-standard model (BSM) physics searches [1]. DUNE will make use of massive liquid argon time projection chamber (LArTPC) detectors and the world's most intense neutrino beam in order to achieve these physics goals. This chapter first explains the workings of LArTPC technology before summarising the three main components of DUNE: the beam production facility, the far detector (FD) and the near detector (ND). Finally, this chapter discusses the Precision Reaction Independent Spectrum Measurement (PRISM) and the data used in the physics analysis presented in this thesis.

3.1 Liquid Argon Time Projection Chamber

The LArTPC is the primary detector technology for DUNE. High resolution images of neutrino interactions, the ability to scale to high detector masses and precise calorimetry make LArTPCs an ideal detector technology for neutrino physics. The first large-scale LArTPC was the ICARUS detector built at Grand Sasso National Laboratory [2, 3]. Figure 3.1 illustrates the operating principles of a typical LArTPC.

A LArTPC consists of a large volume of purified liquid argon within a cryostat, which keeps the argon at a constant temperature. A uniform electric field is applied across the active volume by an anode and cathode plane. Charged particles moving through the detector volume ionise the argon atoms, freeing atomic electrons from

C. M. Hasnip, *DUNE-PRISM—A New Method to Measure Neutrino Oscillations*,
Springer Theses, https://doi.org/10.1007/978-3-032-07251-1_3

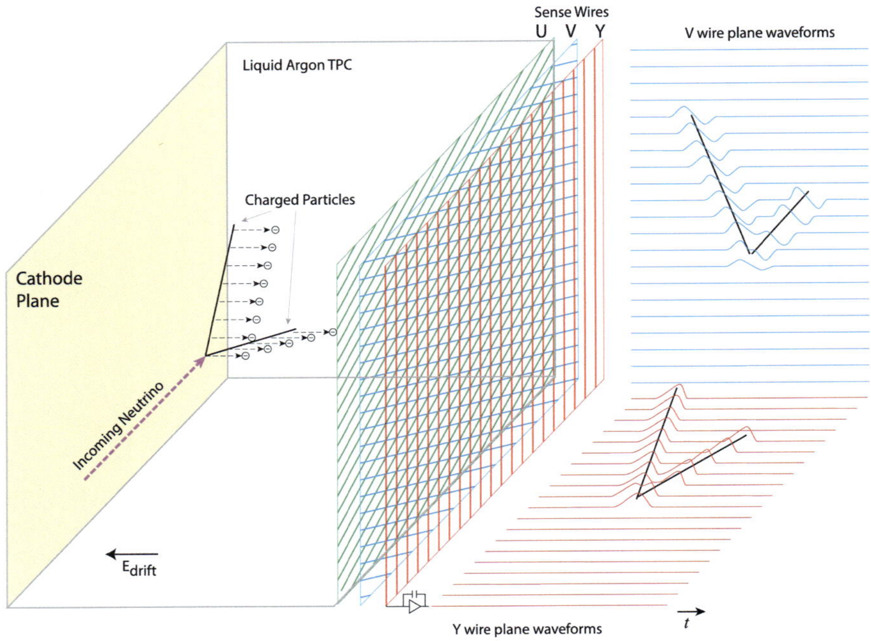

Fig. 3.1 Schematic diagram of a LArTPC [4]. The diagram shows a LArTPC with a horizontal electron drift direction and wire-plane electronic readout design.

their bound state. These electrons are drifted by the field towards the anode plane. The chemically inert nature of argon allows ionisation electrons to drift large distances through the detector without being absorbed. Traditionally, the electric field is applied such that the electrons drift horizontally toward several wire planes, as was the case in the ICARUS detector. However, other LArTPC design choices are available, such as drifting the electrons in the vertical direction [5] and using a pixel-based electronic readout instead of a series of wire planes [6].

In addition to the ionised electrons, liquid argon produces copious amounts of scintillation light during the excitation and ionisation of argon atoms in a neutrino interaction. At zero electric field liquid argon produces approximately 40,000 photons per MeV of energy deposited in the vacuum ultraviolet (VUV) frequency range [7]. Conveniently, liquid argon is transparent to its own scintillation light, allowing the VUV photons to reach the photo-detectors placed behind the anode well before the drifting ionised electrons. Scintillation light in a LArTPC is used to trigger for the start of an event and record the time an event took place, which allows the electron drift time to be independently calculated. In addition, photon detection in a LArTPC can provide a complementary calorimetric energy measurement of an event.

In the traditional LArTPC design, the anode consists of three wire planes—2 induction planes angled at approximately 60° relative to each other and a vertical collection plane—which record the position of the particle track when the ionisation electrons induce an electrical signal on the wires. Electrons pass through the first 2 induction planes and are collected on the third wire plane [8]. Each of the three wire planes provides a two-dimensional image of the event. One dimension is the wire number on the plane. The second dimension is provided by the drift time of the electrons. The drift time is determined by knowing the time at which the event occurred in the detector. For beam events, this can be provided by precise knowledge of the neutrino beam timing. However, for events of interest that do not come from a beam, such as a potential proton decay event, timing information can be provided by scintillation light. The two-dimensional images from the three wire planes are combined to build a three-dimensional image of the event.

3.1.1 Particle Interactions in Liquid Argon

Neutrinos cannot be directly measured by LArTPCs because they are neutral particles that do not ionise the argon atoms. Instead, liquid argon neutrino detectors rely on measuring charged final state particles produced in a neutrino interaction to estimate the initial neutrino's energy. Final state neutral particles, such as neutrons, are difficult to measure directly in a LArTPC. Neutrons can travel large distances in the detector before depositing some portion of their energy and may exit the detector entirely, making it challenging to incorporate neutron energy deposits into the reconstructed neutrino energy. The charged particle interactions can be broadly categorised as either tracks or showers and examples of how these types of interactions appear in a LArTPC event display are shown in Fig. 3.2. LArTPC event displays use a colour gradient to indicate amount of charge deposited on one of the wire planes, where

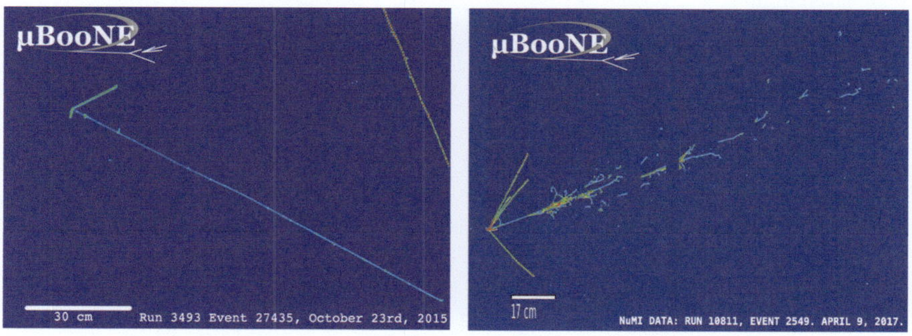

(a) Track **(b)** Shower

Fig. 3.2 LArTPC event displays of neutrino events in the MicroBooNE detector [9]. The muon track is the longest of the three tracks originating from the interaction vertex in Fig. 3.2a. The electromagnetic shower in Fig. 3.2b is produced at the interaction vertex, resulting in many discrete charge depositions across the length of the event display the x-axis is the wire number and the y-axis is the time at which the drifting ionised electrons induced the signal in the wires. The number of ionisation electrons—and therefore the strength of signal induced in the wire planes—corresponds to the amount of energy deposited by a charged particle.

3.1.1.1 Charged Particle Tracks

Tracks are most commonly produced in liquid argon by muons, charged pions and protons. An example of a muon-like track can be seen in Fig. 3.2a. In neutrino experiments the energies of these particles are typically in the range of tens to thousands of MeV. Energy deposits in this energy range are well described by the Bethe-Bloch equation, which calculates the mean energy deposited per unit length, $\langle dE/dx \rangle$ [10]. Figure 3.3 shows the mean energy loss per unit length as a function of muon, pion and proton momenta for several different interaction media according to the Bethe-Bloch equation. A particle with momentum at the minimum of the Bethe-Bloch curve is known as a minimum ionising particle (MIP). Muons and charged pions are typically produced with momenta in the MIP region, which corresponds to energy deposits of approximately 2 MeV/cm in liquid argon [11]. Particle momenta decreases toward the steeply rising left side of the Bethe-Bloch curve as energy is deposited, resulting in a rapid increase in $\langle dE/dx \rangle$ and an abrupt end to the particle track. This is called the Bragg peak. Protons typically move slower than pions or muons and

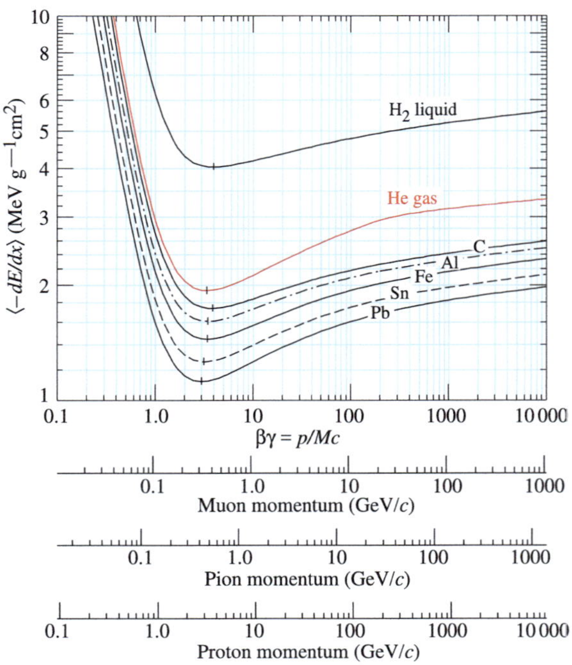

Fig. 3.3 Mean energy loss in several different media as a function of muon, pion and proton momenta according to the Bethe-Bloch equation. Figure taken from Reference [10]

so their tracks end in a Bragg peak close the the neutrino interaction vertex. The Bethe-Bloch equation accounts for the rare events in a track where a large amount of energy is deposited in a single collision, causing an electron, known as a δ-ray, to be knocked out of an argon atom [12].

3.1.1.2 Electromagnetic Showers

At high energies, electron energy loss in matter is dominated by *bremsstrahlung* radiation, whereas photons convert into e^+e^- pairs. This combination of photon radiation and pair production manifests itself in the creation of electromagnetic (EM) showers of cascading electrons, positrons and photons in the detector medium. Particles produced in a neutrino interaction with an argon nucleus that create EM showers are the $e^\pm$, γ and π^0. The neutral pion typically decays as $\pi^0 \to \gamma\gamma$, producing two EM showers. Diagrams illustrating how these EM showers are produced are shown in Fig. 3.4. As the energy of the electrons and positrons diminishes with each radiation length, the energy eventually drops below the critical energy, at which point energy losses are dominated by ionisation. Once these lower energies are reached the EM shower dissipates [10]. An example of an EM shower in the MicroBooNE detector is shown in Fig. 3.2b.

3.1.2 Detector Effects in Liquid Argon

The energy and topology of a neutrino event is reconstructed from the charge observed at the anode from the drifting ionisation electrons. How accurately this process is done is governed by a number of detector effects in LArTPCs. At a typical LArTPC electric field strength of 500 V/cm it can take several milliseconds for electrons to drift to the anode plane, which can be meters from the ionised argon atoms in large-scale TPCs [13]. Not all of the ionised electrons are drifted to the anode; a certain

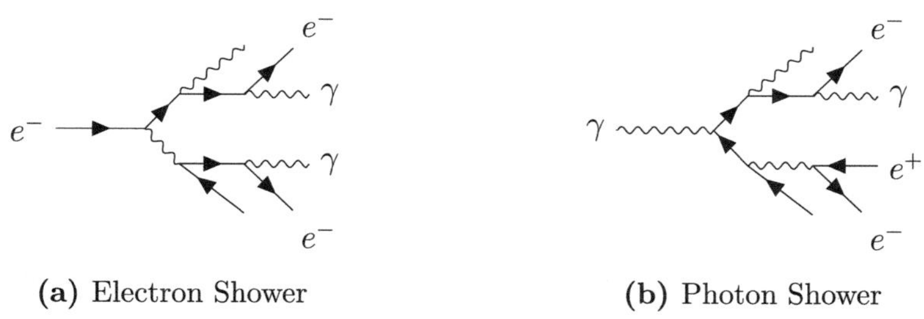

Fig. 3.4 Diagrams illustrating how an electromagnetic shower is produced. Electrons and positrons radiate photons, which subsequently pair produce

proportion recombine with the argon atom to produce scintillation light and the rate at which recombination occurs depends on the strength of the electric field.

Within a large volume of liquid argon there will be impurities, which can capture ionised electrons drifting toward the anode plane. Some of the most problematic impurities are oxygen and water [14, 15]. The effect of these impurities is to suppress the amount of charge observed at the anode plane. The purity of the liquid argon must therefore be carefully monitored when calibrating the detector.

During propagation through the detector volume, the distribution of ionised electron will diffuse over time. The further away from the anode ionisation electrons are produced the more diffusion there will be due to the larger distances travelled through the liquid argon. The presence of the electric field causes the diffusion to be non-isotropic. Electron diffusion is separated into the components transverse and longitudinal to the drift direction. The overall effect of diffusion is to spread out the ionised electrons in space and time, thereby limiting the precision with which the event topology can be measured [16, 17].

Liberating electrons from an argon atom through ionisation also results in an Ar^+ ion, which is drifted by the electric field to the cathode plane. The Ar^+ ions are slower than the electrons, which can lead to a build-up in Ar^+ ions close to the cathode. This distorts the electric field experienced by the drifting electrons, potentially altering the signal measured at the anode. Distortions in the electric field may also affect the rate of recombination, which changes the number of ionised electrons that are drifted to the anode. These are called space charge effects and they are most relevant in detectors close to the Earth's surface where there is a high cosmic ray background [18].

3.1.3 Scintillation Light in Liquid Argon

LArTPCs are equipped with a photo-detection system to measure the scintillation light produced from neutrino interactions on argon. Charged particles moving through liquid argon produce excited argon dimer states, which subsequently de-excite to emit scintillation light. There are two processes by which this can happen:

- An argon atom is excited and couples with an adjacent ground state argon atom to form an excited argon dimer state, which subsequently de-excites through the emission of a photon,

$$\mathrm{Ar}^* + \mathrm{Ar} \rightarrow \mathrm{Ar}_2^* \rightarrow 2\mathrm{Ar} + \gamma. \tag{3.1}$$

- Charged particles ionise an argon atom, which couples to a charge-neutral argon atom to form a Ar_2^+ state. The Ar_2^+ state may then recombine with an ionisation electron to form an excited dimer that similarly decays through photon emission,

$$\mathrm{Ar}_2^+ + e^- \rightarrow \mathrm{Ar}_2^* \rightarrow 2\mathrm{Ar} + \gamma. \tag{3.2}$$

The γ in each case is a VUV photon with a wavelength spectrum peaked at 128 nm. Excited argon dimer states can be in either a singlet or triplet state, which have average lifetimes of approximately 6 ns and 1.6 ns respectively [19]. The strength of the electric field in the TPC, the type of charged particle ionising the argon and the argon purity can all affect the number of photons produced per unit of energy deposited. For example, a higher electric field strength will suppress the amount of scintillation light emitted though ionisation because more electrons will be drifted to the anode plane before recombination can occur. Approximately 24,000 photons are expected to be emitted per MeV of deposited energy while operating at a typical field strength of 500 V/cm [4].

3.2 Producing a Neutrino Beam for DUNE

Measuring neutrino oscillations first requires a reliable and intense source of neutrinos. The DUNE neutrino beam is provided by the long baseline neutrino facility (LBNF), based at Fermi National Accelerator Laboratory (Fermilab). The first step in producing the LBNF beam is to accelerate H^- ions through the Fermilab Linac up to 400 MeV. The hydrogen ions are subsequently stripped of their electrons to produce a proton beam that is accelerated up to 8 GeV by the Booster proton synchrotron. Finally, the protons enter the Fermilab Main Injector, which is a larger synchrotron that can accelerate the protons to energies between 60 GeV and 120 GeV. Protons enter the Main Injector in "spills" lasting 10 μs. Each spill is divided into 6 "batches" formed of 84 discrete bunches, of which 81 are filled with protons. The total number of protons in each spill is approximately 7.5×10^{13} and all of the protons in the spill are extracted from the Main Injector by a kicker magnet every 1.2 s.

After acceleration, each spill of protons is directed out of the Main Injector and into the Primary Beam Encloser that transports the high-energy protons to the LBNF target hall. The initial beam power is 1.2 MW, rising to 2.4 MW after several years under the Proton Improvement Plan II (PIP-II) upgrade [21]. PIP-II is broad program of improvements to the accelerator facilities at Fermilab, including upgrading the Linac to accelerate the hydrogen ions to 800 MeV. With 120 GeV protons and a 1.2 MW beam, 1.1×10^{21} protons-on-target (POT) are expected each year. Figure 3.5 shows a schematic diagram of LBNF from the Main Injector to the near detector site.

The target hall design has been optimised to meet the DUNE science goals, with particular emphasis on assisting the discovery of CP violation in the neutrino sector [22]. Upon entering the target hall, the protons pass through a hole in a 1.5 m hollow graphite cylinder called the baffle, which prevents damage to downstream facilities by any misalignment in the proton beam and allows aligned protons to pass through unimpeded. As discussed briefly in Sect. 2.4.4, protons subsequently interact in the graphite target to produce hadrons, a large fraction of which are charged pions. The optimised LBNF design then uses three separate magnetic horns to focus the charged hadrons into a collimated beam. The magnetic horns are configured to either focus the π^+ and de-focus the π^-, called forward horn current (FHC), or to

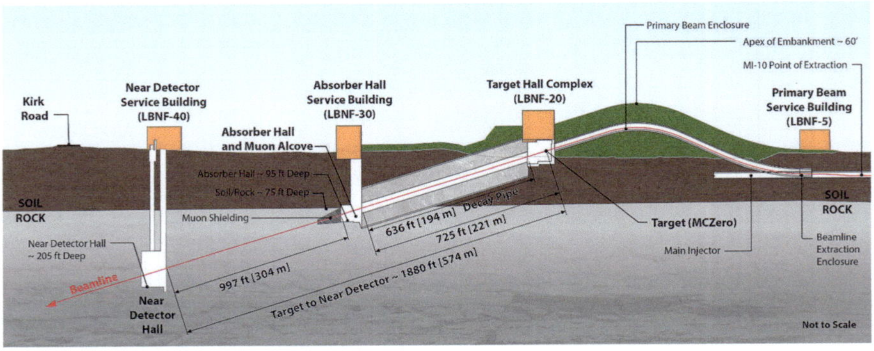

Fig. 3.5 Schematic diagram of the long baseline neutrino facility (LBNF). Figure from Ref. [20]

focus the π^- and de-focus the π^+, called reverse horn current (RHC). After focusing, the hadrons enter the helium-filled decay pipe, where most of the hadrons decay. Practically all of the charged pions decay as $\pi^+ \rightarrow \nu_\mu \mu^+$ and $\pi^- \rightarrow \bar{\nu}_\mu \mu^-$. Hence the neutrino beam is enriched with either ν_μ or $\bar{\nu}_\mu$ neutrinos, depending on whether the magnetic focusing horns are configured in FHC or RHC mode respectively. An absorber at the end of the decay pipe, constructed from aluminium, steel and concrete, stops any remaining hadrons from continuing to the ND hall. Further details on the LBNF design can be found in References [20] and [23] and a cartoon illustrating a simplified version of the LBNF target hall and decay pipe is shown in Fig. 3.6.

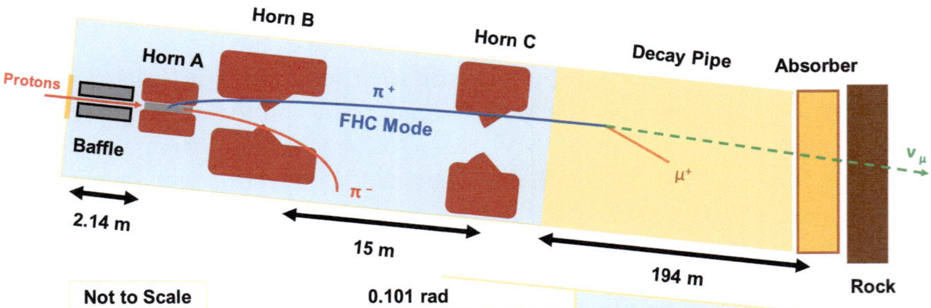

Fig. 3.6 Illustration the LBNF target hall and decay pipe. Protons enter from the left through a beryllium "window" before passing through the baffle. Protons interact with the target that is surrounded by Horn A. Hadrons are subsequently focused by Horn B and C before entering the decay pipe. The proton beam is tilted down by 0.101 radians. A hadron absorber stops any remaining hadrons in the beamline

3.3 DUNE Far Detector

The neutrino beam produced by LBNF will travel 1285 km through the Earth's crust to the Sandford Underground Research Facility (SURF), which is a mile underground. This is the location of the DUNE FD. The FD will be a facility of four detector modules, each with an active volume mass of 17 kt and a fiducial mass of 10 kt, giving a total detector fiducial mass of 40 kt [24]. Installation of the four modules will be staged over a number of years, beginning with two modules for the initial period of data-taking. The primary technology of choice for the detector modules is the LArTPC, although there are on-going discussions on whether the fourth module (FD4) could use a different technology. At least one of the detector modules will use a horizontal-drift single-phase LArTPC design [25]. An illustration of a FD module cryostat is shown Fig. 3.7. The second FD module (FD2) will be a single-phase vertical drift configuration. A single-phase configuration has the electrons drift only through liquid argon. This is in contrast to a dual-phase design, often employed for dark matter detectors such as LUX-ZEPLIN [26], where a gas region close to the anode plane amplifies the signal. However, the excellent argon purity and electron drift distances achieved by ProtoDUNE Single-Phase and Dual-Phase renders the amplification from a gas phase unnecessary for DUNE [5].

3.3.1 Horizontal Drift Module

The horizontal drift module will be the first module to be installed at the FD site. It is a traditional LArTPC design, successfully deployed in MicroBooNE [4] and

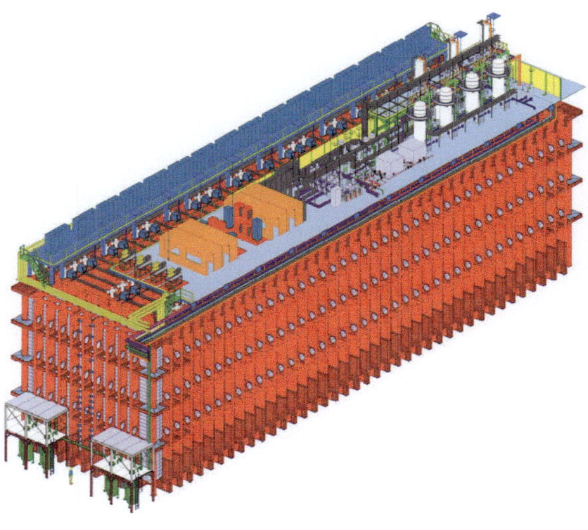

Fig. 3.7 Schematic diagram of the FD detector module from Ref. [25]

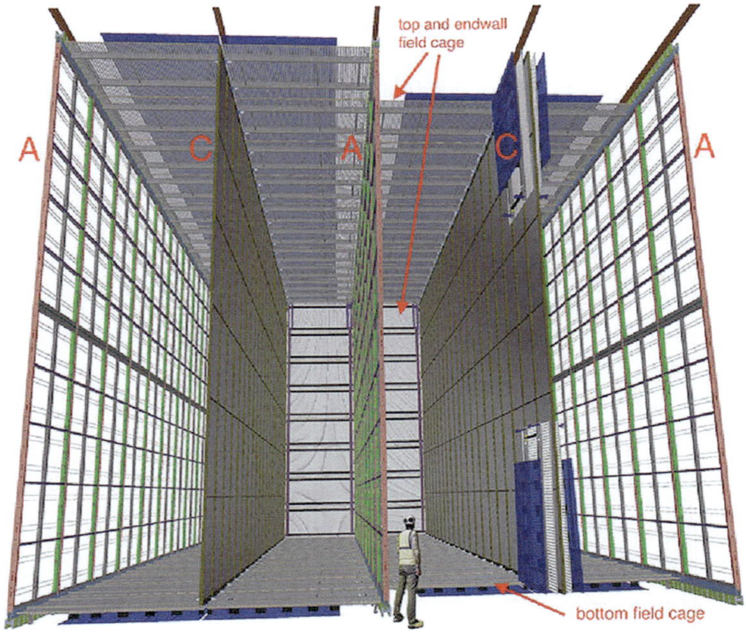

Fig. 3.8 Schematic diagram of the FD horizontal drift TPC from Ref. [25]

ProtoDUNE Single-Phase [13] and described extensively in Sect. 3.1. Further details on the horizontal drift module design can be found in Reference [25]. The active liquid argon volume of the module is 58 m long, 12 m high and 13 m wide, where the width is separated into four 3.5 m drift volumes bounded by an anode and a cathode wall. The geometry of the horizontal drift active volume, including the four drift volumes and the anode and cathode walls, is shown in Fig. 3.8.

Each cathode wall is made from an array of cathode plane assemblies (CPAs). An array of CPAs is formed from 25 adjacent CPA planes that span the full-height of the FD module. The CPA planes are made of two panels, which in turn are constructed from three units of resistive panels. The surface of each CPA is a highly resistive material, which in the case of a sudden breakdown of the high voltage prevents electrical discharges from damaging the detector components. Each CPA array is biased at -180 kV, whilst the anode walls are held at ground. This provides a uniform electric field of 500 V/cm across each drift volume. The field cage running around the active argon volume assists in keeping the electric field constant across the drift region.

The anode wall is an array of anode plane assemblies (APAs). Each APA is 6 m tall and 2.3 m wide, requiring 50 APAs to cover one anode wall with two rows of adjacent APAs stacked on top of one another. The APA is covered by three planes of active wires: the X plane is vertical and the U and V planes are angled at $\pm 35.7°$ relative to the X plane. The wire planes are wrapped around the APA and electronic

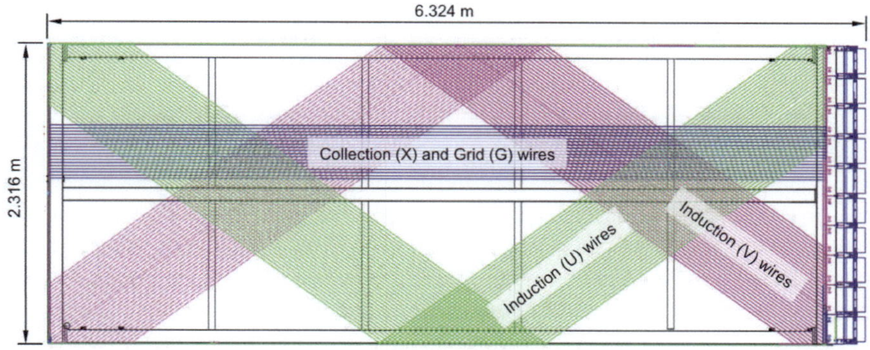

Fig. 3.9 Schematic diagram of the horizontal drift model APA frame from Ref. [25]. The different wire planes are drawn in addition to the cold electronics boards highlighted in blue on the right edge of the APA

signals induced by drifting electrons are readout by the front-end electronics at one side of the APA, as illustrated in Fig. 3.9. An additional plane of wires G is used for shielding and is not readout by the electronics. The front-end electronics are held in the liquid argon with the APA frame, taking advantage of the low temperatures to improve the gain and reduce the electronic noise in amplifying and digitising the electronic signals from the wire planes.

Photon detection systems are placed in the inactive region between the wire planes of the APAs, so as to maintain the entire active volume for drifting electrons. As discussed in Sect. 3.1, the photon detection system is used to record the time an event occurred, in addition to complementary calorimetric energy measurements. The confined space between the APA wires necessitates the use of silicon photomultipliers (SiPMs) as the photon detectors. SiPMs have a relatively small detection area, motivating the employment of a light collection system in order to economise the number of SiPMs required to achieve the necessary coverage for photon detection across the FD module. The FD horizontal drift module will use the X-ARAPUCA system. A development of the original ARAPUCA design [27], the X-ARAPUCA system traps wavelength shifted photons inside a box containing one or several SiPMs. The internal walls of the box are highly reflective, allowing the photons to be contained until they are detected by the SiPMs [28].

3.3.2 Vertical Drift Module

The horizontal drift module design to be used in the first FD module employs well-established LArTPC technologies. Research and development is on-going into new LArTPC detector designs that could improve the performance and reduce the cost and complexity of the other FD modules. The default design currently under development for the second FD module (FD2) is the vertical drift single-phase module [5]. In the vertical drift FD module, the active volume of liquid argon is separated into two 6.5 m vertical drift regions stacked on top of one another and separated by a single cathode wall across the centre of the active volume. There are two anode walls, one at the top of the detector above the cathode and one at the bottom of the detector below the cathode. The longer drift distances and detector components close to the cryostat walls maximises the volume available for detecting neutrino interactions. The anode wall is constructed from many perforated printed circuit boards. The printed circuit boards perform the role of the wire planes in the horizontal drift design, but are generally more robust and can be mass-manufactured. The vertical drift design is a very promising technology and will be tested in the next iteration of the ProtoDUNE detectors at CERN.

3.3.3 Data Acquisition and Triggering

The FD data acquisition (DAQ) system records and processes data from the DUNE FD modules. In addition, the DAQ provides timing information and synchronises the different detector modules. The DAQ system can be separated into the upstream, data selection and back-end components. In the case of the horizontal drift module discussed in Sect. 3.3.1, the front-end electronics positioned at the end of each APA frame are readout by the upstream DAQ. The upstream DAQ design is based on the FELIX system built for the LHC [29].

Data from the upstream DAQ is fed to the data selection subsystem, which is responsible for making triggering decisions. There are two types of triggering system for the FD modules. The localised high-energy trigger opens a readout window of 5.4 ms for beam, cosmic and nucleon decay events, which has an efficiency greater than 99% at 100 MeV of deposited energy. A second trigger system opens an extended readout window of 10 s to search for low-energy deposits from a supernova neutrino burst (SNB). Once triggering decisions are made, the recorded data is buffered until it is ready to be received by the back-end DAQ, which transfers the selected data to permanent storage. Further details on the DUNE FD DAQ and triggering system can be found in Reference [25].

3.3.4 Calibration

The DUNE FD will be taking data over many years. As such, it is vital that the detector response of the large FD modules is stable and well-understood over the lifetime of the experiment. This is achieved through calibration. Calibration is required to understand aspects of the detector response, such as the energy scale of leptons and hadrons, vertex position resolution and the fiducial volume size. Knowledge of the detector response is achieved by measuring fundamental detector performance parameters, such as the energy loss per unit length, the electric field uniformity and the electron drift lifetime. A common calibration method used for LArTPCs is to measure the energy loss per unit length of cosmic ray muons [30]. However, at the depth of the DUNE FD the cosmic muon rate is very low, making it challenging to rely this calibration technique without input from other methods.

DUNE will employ a variety of techniques to calibrate the FD modules in different energy regimes. Neutron emitters and radioactive sources, such as the ^{39}Ar distributed across the argon volume, provide low energy calibration sources. At energies of the order 100 MeV, Michel electrons, π^0 decays, stopping protons and muons can be used as calibration sources, originating first from cosmic ray events and eventually from the LBNF beam. In addition, a detector monitoring system will be in place that tracks the purity and temperature of the argon and a laser will be used to map the electric field – a system already implemented in MicroBooNE [31]. Plans for the FD calibration are described in more detail in Reference [25].

3.4 DUNE Near Detector Complex

DUNE will be equipped with a highly capable near detector (ND) situated 575 m downstream from the neutrino beam target. Figure 3.10 shows a schematic diagram of the ND complex, which is composed of three separate detector components. The first component is a LArTPC called ND-LAr, which has a modular design and acts as the primary target for neutrino interactions. Downstream from ND-LAr is a tracking

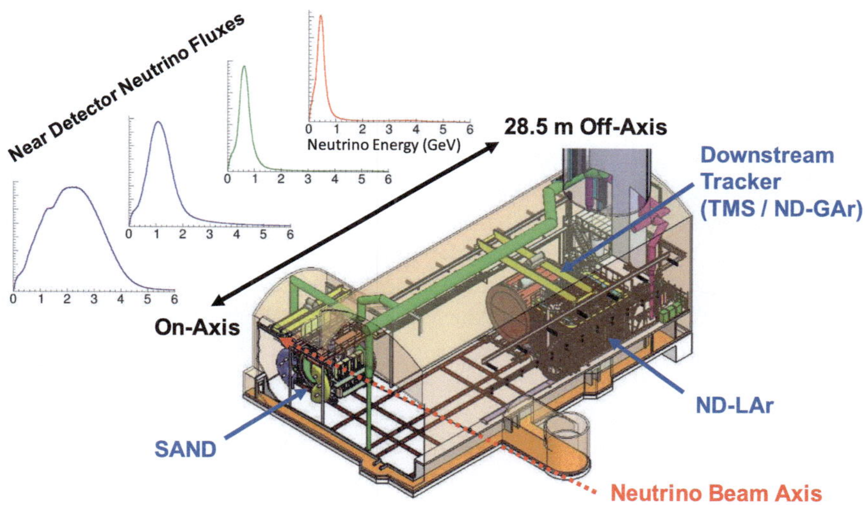

Fig. 3.10 Schematic diagram of the ND detector complex. The diagram shows ND-LAr and ND-GAr in the far off-axis position at 28.5 m from the beam axis centre [33]

detector capable of measuring the momentum of muons exiting the back of ND-LAr. In the first phase of DUNE this downstream tracker will be the Temporary Muon Spectrometer (TMS) and in the second phase it will be replaced by a pressurised argon gas TPC called ND-GAr. In addition to measuring muons exiting ND-LAr, ND-GAr will be able to measure its own sample of neutrino interactions on argon. While the FD is positioned on the axis of the neutrino beam, ND-LAr and TMS/ND-GAr will be able to move horizontally up to 28.5 m off the axis of the neutrino beam. This feature of the ND is called the DUNE Precision Reaction Independent Spectrum Measurement (DUNE-PRISM). Since ND-LAr and TMS/ND-GAr will spend significant amounts of time off-axis, the System for on-Axis Neutrino Detection (SAND) stays in the on-axis position and monitors the stability of the neutrino beam. Further details on all the components of the DUNE ND complex can be found in Ref. [32].

3.4.1 Near Detector Liquid Argon

The choice of liquid argon as the primary detector media at the DUNE FD necessitates the inclusion of a liquid argon component at the ND. The ND can better control the impact of cross section and detector uncertainties on the oscillation analysis if the target nucleus for neutrino interactions and detector technology is the same as the FD. The ND-LAr has 67 t fiducial mass contained in an active volume that is 7 m wide perpendicular to the beam direction, 5 m long in the direction of the beam and 3 m high. The ND is situated close to the LBNF target hall, resulting in very high neutrino interaction rates. This would be a challenge for a typical LArTPC, since the speed at which the detector can record neutrino events is limited by drift time for the ionised electron in liquid argon, which is typically a few milliseconds. Consequently, the use at the ND of a traditional LArTPC design, such as those seen in MicroBooNE [4] and SBND [34], could result in unacceptable levels of event pile-up. The DUNE collaboration has therefore developed a modular LArTPC design for ND-LAr that splits the active volume of liquid argon into many smaller drift regions.

The design of ND-LAr is based on the ArgonCube concept [35]. It consists of 35 optically segregated modules that act as independent LArTPCs. Each module spans the full 3 m height of the detector and the width of ND-LAr is 7 modules wide and 5 modules long, as can be seen in Fig. 3.11. Segmenting ND-LAr into TPC modules with shorter electron drift distances reduces the necessary high voltage for the electric field and the required argon purity level.

In each module there are two 3 m high anode walls on each side of the TPC, which is segmented by a single cathode wall. This creates two drift volumes in each ND-LAr module with a 50 cm electron drift distance. The anode planes use a pixelated charge readout system called LArPix [36], which significantly reduces the complexity of three-dimensional image reconstruction compared to the wire plane APAs.

The scintillation light is of great importance to the ArgonCube design. As well as providing an independent measurement for the start of a neutrino event in a module,

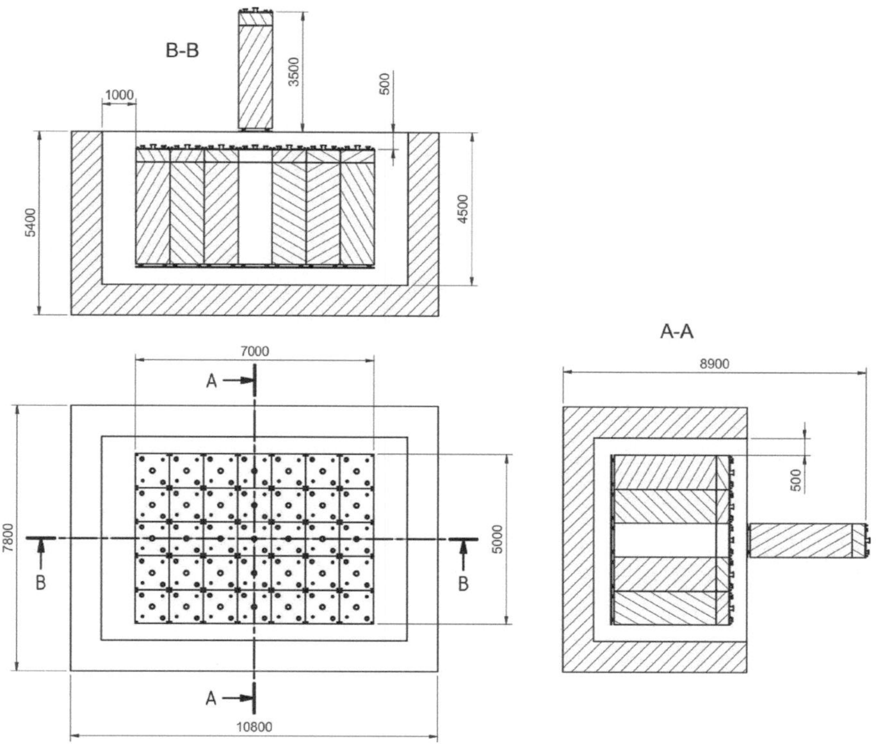

Fig. 3.11 Schematic diagram of the ND-LAr module layout

it is also used to reject charge depositions from the event reconstruction that are not associated with the event of interest. This mitigates the impact of pile-up on the event reconstruction. The light collection system is mounted behind the field cage on the sides of the module that do not hold the anode walls. The technologies used in the light collection system are the light collection module (LCM) and ArCLight [37].

3.4.2 Downstream Tracker

A significant fraction of the μ^- (μ^+) produced in ND-LAr when a ν_μ ($\bar{\nu}_\mu$) interacts in the liquid argon will exit out the back of ND-LAr. Since LArTPCs measure the energy of a particle by the range of its track, most high-energy exiting muons would not have their energy accurately reconstructed by ND-LAr. The purpose of the downstream tracker is to measure the energy of these exiting muons with high-precision.

The first phase of DUNE will install the Temporary Muon Spectrometer (TMS) behind ND-LAr. TMS is a magnetised iron and scintillator plane tracker similar in design to the MINERνA detector [38]. The sign of the muon, and therefore whether the interacting neutrino was a ν_μ or $\bar{\nu}_\mu$, is determined by the magnetic field. The momentum of the muons exiting ND-LAr and entering TMS is measured by the range of the muon track.

After several years of data taking, TMS will be replaced by a magnetised high-pressure argon gas TPC called ND-GAr. The functioning principle of the gas TPC is similar to the LArTPC, in that charged particles ionise the argon atoms in the gas and the resulting electrons are drifted by an electric field to the readout system. ND-GAr will use the existing readout chambers from the ALICE detector at CERN [39]. Since ND-GAr is magnetised, it acts as a muon spectrometer that distinguishes between ν_μ and $\bar{\nu}_\mu$ events. The muon momentum is measured by the curvature of the track in the magnetic field. ND-GAr also measures an independent sample of neutrino interactions on argon with very low particle tracking thresholds. The argon gas TPC will be able to resolve low energy protons and pions close to the neutrino interaction vertex, which will enhance the ND's ability to constrain the systematic uncertainty in the neutrino interaction model [40].

3.4.3 SAND

The neutrino flux model parameterises the design features of the LBNF, such as the position of the magnetic focusing horns, the dimensions of the decay pipe and the density of the graphite target. Each of the parameters that control these features in the flux model have an assumed central value and a corresponding systematic uncertainty. Over the lifetime of the experiment the design features of LBNF may change in unexpected ways. For example, the density of the graphite target could degrade earlier that expected. Whilst hardware will be in place to monitor some these parameters at the LBNF target hall, it is preferable to have a means of monitoring changes in the neutrino flux directly by making neutrino flux measurements at the ND. This is the role of SAND, which primarily acts as a neutrino flux monitor that is placed in the centre of the neutrino beam axis, as shown in Fig. 3.10. As will be discussed in the next section, the fact that ND-LAr and TMS/ND-GAr can move off the axis of the neutrino beam makes it all the more important that a detector is always available on-axis to monitor and diagnose changes in the neutrino flux (Fig. 3.12).

The design of SAND is centred around the refurbishment of the KLOE experiment's superconducting magnet and electromagnetic calorimeter [42, 43]. The KLOE magnet provides a 0.6 T field in a cylinder of 4.8 m diameter and 4.3 m length. The electromagnetic calorimeter is constructed from lead and scintillating fibre. Additional components of SAND are the inner tracker and a potential liquid argon component called GRanular Argon for Interaction of Neutrinos (GRAIN). The inner tracker design is a series of orthogonal straw tube trackers that provide both additional particle identification capabilities and a source of carbon nuclear targets for neutrino interactions. ND-LAr and ND-GAr are both argon targets, therefore the additional nuclear targets present in SAND can assist in constraining neutrino cross section uncertainties.

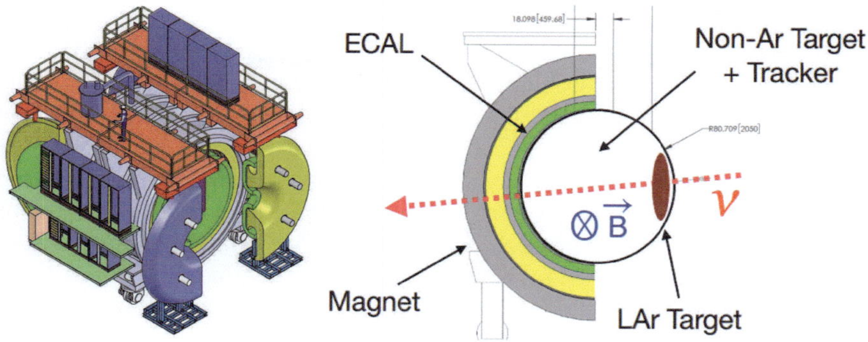

Fig. 3.12 Schematic diagram of SAND. Figure created and copyright owned by Matteo Vicenzi [41]

3.5 PRISM

As stated at the beginning of Sect. 3.4, ND-LAr and TMS/ND-GAr are capable of moving horizontally by up to 28.5 m perpendicular to the neutrino beam axis. This mechanism is called Precision Reaction Independent Spectrum Measurement (PRISM) and is the focus of this thesis. The DUNE neutrino flux is dominated by the two-body decay of charged pions in the LBNF target hall. The kinematics of this two-body decay results in a neutrino flux that becomes increasingly monochromatic as the angle relative to the beam axis increases. Figure 3.13 demonstrates this by showing how the relationship between the neutrino energy and the energy of the parent pion becomes increasingly flat as the angle increases. Likewise, Fig. 3.13 also shows the neutrino flux shape narrowing and moving to lower energies as the angle

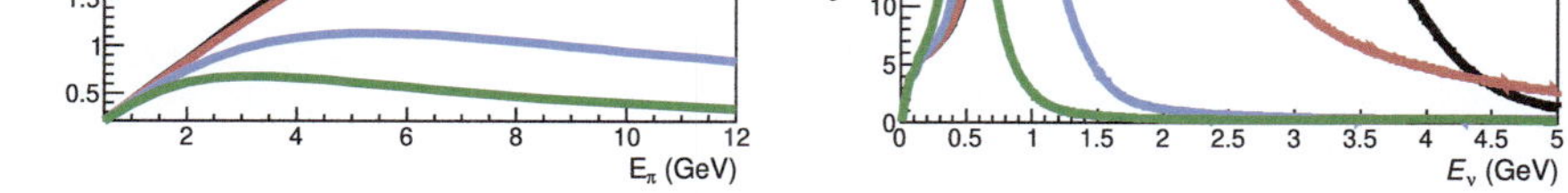

Fig. 3.13 Pions are produced when protons hit the LBNF target. On the left is the neutrino energy as a function of the pion energy for different angles relative to the neutrino beam axis. The right plot shows the corresponding neutrino flux spectra for the same four angles. At high angles the neutrino energy becomes relatively constant with pion energy, resulting in an approximately monochromatic neutrino flux

relative to the beam axis increases. This phenomena is used in the T2K and NOνA experiments to achieve a relatively monochromatic flux peaked at the neutrino energy for which the oscillation probability is expected to be maximal [44, 45].

DUNE is an on-axis long baseline oscillation experiment, meaning the ND and FD are positioned in the centre of the neutrino beam axis. However, PRISM uses the relationship between the neutrino flux and the off-axis position relative to the beam axis to perform a novel data-driven neutrino oscillation analysis. This type of data-driven analysis was first proposed by the nuPRISM Collaboration for the T2K and Hyper-K experiments [46]. The ND measures different neutrino fluxes when taking data at off-axis positions. Many measurements of off-axis neutrino events can be linearly combined to produce an event rate spectrum that closely matches the oscillated spectrum measured in the FD. This data-driven prediction can then be fitted to the measured FD event rate to extract the oscillation parameters of interest.

PRISM presents a very different approach to the traditional oscillation analysis methodology, where the ND is used to constrain the parameters of the flux, neutrino cross section and detector models. The MC prediction of the FD event rate is then made using the models constrained by the ND data. However, no complete model for the neutrino interactions with nuclei exists and the difficulties in modelling these interactions was discussed at length in Sect. 2.6. The PRISM approach is to build the FD event rate prediction directly from the ND data, which does contain the correct neutrino interaction physics. Building a prediction of the FD event rate from ND data allows the oscillation parameters to be extracted in a way that naturally includes any unknown or poorly modelled cross section effects. There are numerous ways off-axis measurements at the ND can enhance the capabilities of DUNE and more details can be found in Reference [32]. However, this thesis will focus on the development of a comprehensive PRISM oscillation analysis.

3.6 Monte Carlo Data

The physics analysis presented in this thesis relies on simulated Monte Carlo (MC) data produced by the DUNE Collaboration. A neutrino oscillation analysis does not require all the detector response information, therefore the MC data for the near and far detectors is truncated into Common Analysis Format (CAF) files, which maintain only the high-level neutrino event information relevant to a physics analysis. The MC CAF files provide true and reconstructed kinematic information for ν_μ, ν_e, $\bar{\nu}_\mu$ and $\bar{\nu}_e$ signal events in the ND and FD for the two different horn polarities (FHC and RHC) and background events.

In order to produce the CAF files, simulation and reconstruction of neutrino events is performed in the near and far detectors. Neutrino events are simulated in the near and far detectors using a common flux simulation tool called G4LBNF, which employs a detailed GEANT4 simulation of the LBNF facility to produce neutrino fluxes for each neutrino flavour and the two horn polarities at the ND and FD [20]. Figure 3.14 shows the G4LBNF-generated neutrino fluxes in the ND and

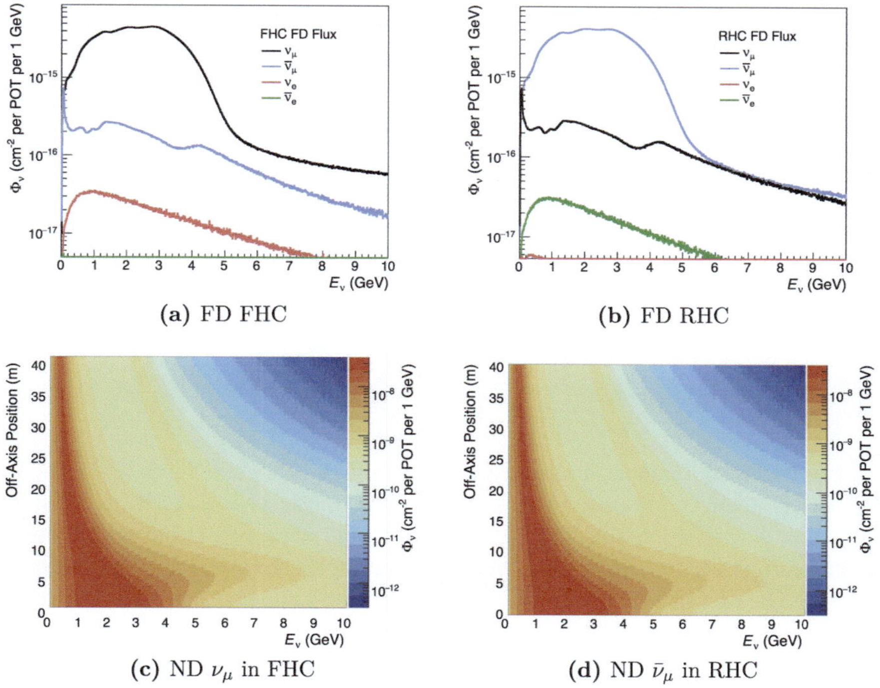

(a) FD FHC

(b) FD RHC

(c) ND ν_μ in FHC

(d) ND $\bar{\nu}_\mu$ in RHC

Fig. 3.14 Fluxes simulated in the FD and ND by G4LBNF for different neutrino flavours. FD fluxes in neutrino enhanced and anti-neutrino enhance mode for ν_μ, $\bar{\nu}_\mu$, ν_e and $\bar{\nu}_e$ fluxes shown in Fig. 3.14a, b. G4LBNF generates neutrino fluxes at the ND both on-axis and going up to 40 m off-axis. Shown are the ν_μ and $\bar{\nu}_\mu$ fluxes generated at all off-axis positions in FHC and RHC mode respectively

FD. The Generates Events for Neutrino Interaction Experiments (GENIE) generator is used to simulate neutrino interactions on argon nuclei in the FD and ND [47]. The GENIE version is GENIE 2.12.10. Full reconstruction of the final state particles is performed on the simulated neutrino interactions in the FD. However, at the time of writing no detector simulation or reconstruction is available in the ND. Therefore, parameterised reconstruction is used for the ND MC data.

3.6.1 Far Detector Simulation

Each DUNE FD module is 10 kt. However, it would computationally expensive to attempt to simulate neutrino interactions in the entire volume. Therefore, a smaller FD module geometry is simulated, which is 13.9 m long, 12.0 m high and 13.3 m wide. For this analysis all four FD modules are assumed to use the horizontal drift design described in Sect. 3.3.1. G4LBNF generates a neutrino flux in this FD geometry for

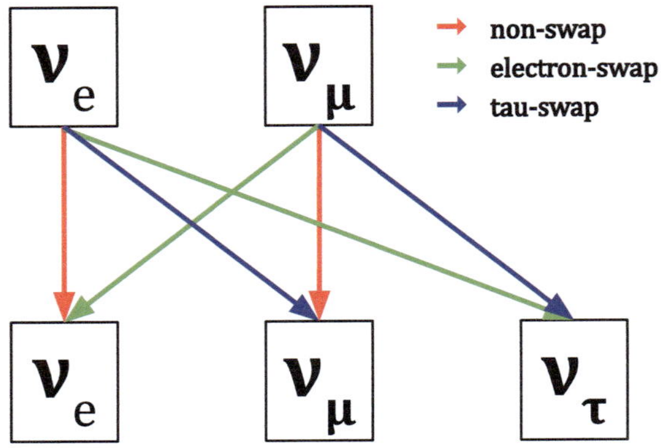

Fig. 3.15 Illustration of the three types of FD flux files that allow for three-flavour oscillations for any set of parameters. The "non-swap" files do not change any neutrino flavour, whilst the "electron-swap" and "tau-swap" do change the initial neutrino flavours

FHC and RHC horn configurations, as shown in Figs. 3.14a and 3.14b respectively. The LBNF beam is dominated by either ν_μ in FHC mode, or $\bar{\nu}_\mu$ in RHC, in addition to smaller ν_e and $\bar{\nu}_e$ components. It is assumed ν_τ and $\bar{\nu}_\tau$ are negligible contributions to the beam at production. Three separate flux samples are produced for each horn polarity. The first sample ("non-swap") maintains the same neutrino flavour composition as the nominal LBNF flux. The second sample ("electron-swap") converts the flavour of the ν_μ ($\bar{\nu}_\mu$) flux to ν_e ($\bar{\nu}_e$) and the flavour of the intrinsic ν_e ($\bar{\nu}_e$) flux into ν_τ ($\bar{\nu}_\tau$). Finally, the third sample ("tau-swap") converts the flavour of the ν_μ (ν_μ) flux to ν_τ ($\bar{\nu}_\tau$) and the flavour of the intrinsic ν_e ($\bar{\nu}_e$) flux into ν_μ ($\bar{\nu}_\mu$). Figure 3.15 illustrates how the three flux sample types swap the neutrino flavours. The three types of flux sample allow oscillation probabilities to be applied to the flux of any neutrino flavour and for any oscillation parameter hypothesis. Neutrino events in the FD can then be generated without assuming the value of any oscillation parameters.

GENIE generates neutrino interactions on argon nuclei inside the FD active volume for each of the flux samples. A GEANT4 simulation of the FD geometry then propagates the final state particles produced by GENIE through the FD volume. There are six FD MC CAF files for the three different flux samples and two horn polarities. Neutrino interactions are generated in the FD active volume with exposures ranging from 10^{24}–10^{25} POT, which corresponds to several thousand years of data taking in the LBNF beam. Many more MC events are generated than will be measured by the DUNE FD in order to suppress statistical fluctuations in the FD MC data.

Final state particles generated by GENIE produce ionisation electrons and scintillation light that induce electronic signals on the wire planes and photo-detectors respectively. The GEANT4 particle propagation, wire plane readout and signal digitisation are all simulated by the LArSoft package, which has been validated by numerous other LArTPC experiments, such as MicroBooNE and ProtoDUNE [48]. The ionisation electrons on each wire are converted to raw wire signals through a convolution of the simulated field and electronics response, resulting in a series of digitised waveforms on each wire that are a function of analog-to-digital converter (ADC) units and time.

3.6.2 Far Detector Reconstruction

The LArSoft package performs the first step in event reconstruction, which is the deconvolution of the raw wire signals. Deconvolution removes the effect of the LArTPC electric field and electronic response to provide a standard Gaussian waveform for each wire signal, the size of which is determined by the charge induced in the wire. An algorithm then scans the waveform for each wire to find local maxima and minima that meet a minimum charge threshold. If the threshold is reached then the signal induced on the wire is called a "hit". The time coordinate and total charge deposited in each hit is determined by fitting a Gaussian function to each hit, where the area under the Gaussian function is the total charge deposited. The wire planes are calibrated so that the amount of charge deposited in a single hit corresponds to the energy lost by the particle while traversing the liquid argon. At this point, corrections are applied to the energy reconstruction associated with each hit to account for ionisation electron capture on impurities in the argon and recombination. An example of three reconstructed hits in ProtoDUNE Single-Phase is shown in Fig. 3.16.

Once wire hits are reconstructed, clustering algorithms collect nearby wire hits together that occur in close proximity in space and time. The clustering algorithm then assigns these clusters of hits to reconstructed objects, such as tracks or EM showers. The Pandora package provides a multi-algorithm approach for hit clustering for DUNE data and has been successfully employed for analysing ProtoDUNE and MicroBooNE data [49–51].

The reconstructed energy of a CC neutrino interaction is the sum of the primary lepton and hadronic energy, reconstructed by the LArSoft hit finder and Pandora clustering algorithm. A muon produces a long track and an electron results in an EM shower. If the event selection identifies an event as a ν_μ or $\bar{\nu}_\mu$, then the lepton energy is taken from the longest reconstructed track. Tracks that are fully contained within the detector have their energy reconstructed by integrating the measured energy-loss per unit distance (dE/dx) over the length of the track. Tracks that exit the detector

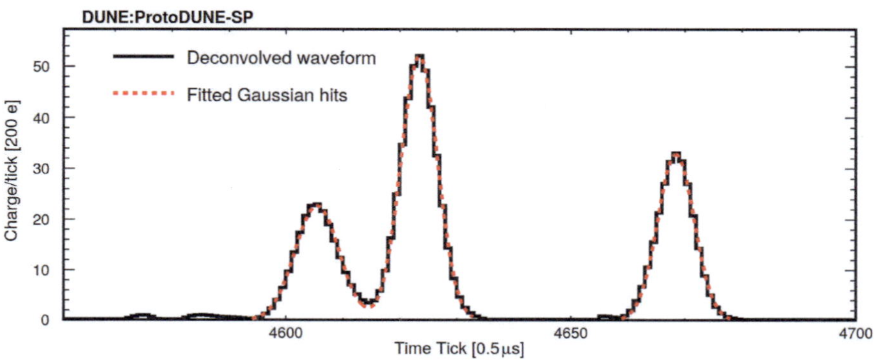

Fig. 3.16 Three reconstructed hits fitted with a Gaussian from the ProtoDUNE Single-Phase. Figure by the DUNE Collaboration [49]. Published by IOP Publishing Ltd on behalf of Sissa Medialab

have their energy estimated by multiple Coulomb scattering. If an event is selected as a ν_e or $\bar{\nu}_e$, the highest energy EM shower is identified as the reconstructed lepton. The hadronic energy is defined as the sum of the energy from all the reconstructed hits not associated with the reconstructed lepton track. Neutral particles, such as neutrons, are difficult to reconstruct and complex final state interactions can occur within the argon nucleus that are not observable. Therefore, the neutrino interaction model is used to apply corrections to the reconstructed hadronic energy to account for this missing energy.

3.6.3 Near Detector Simulation

G4LBNF also generates the neutrino flux at the ND for up to 40 m off the axis of the neutrino beam, as shown in Fig. 3.14c, d. Different flux samples with swapped neutrino flavours are not required in the ND as neutrino oscillation probabilities are negligible at the DUNE ND. GENIE generates neutrino interactions on the argon in ND-LAr, which is simulated by GEANT4. Neutrino interactions are generated with the detector at particular off-axis positions. Table 3.1 shows the different detector

Table 3.1 The POT generated for the ND MC data at each off-axis stop in FHC mode and RHC mode. One year of data taking corresponds to 1.1×10^{21} POT. Therefore, in total there is 6.58 years of FHC ND MC and 7.73 years of RHC ND MC assuming a 1.2 MW beam power

Detector position (m)	FHC mode POT	RHC mode POT
0.00	2.07e+21	3.20e+21
−1.75	2.61e+20	2.79e+20
−2.00	2.89e+20	2.89e+20
−4.00	2.84e+20	2.72e+20
−5.75	2.80e+20	2.80e+20
−8.00	2.89e+20	2.99e+20
−9.75	2.91e+20	2.94e+20
−12.00	2.99e+20	3.10e+20
−13.75	2.99e+20	2.96e+20
−16.00	2.15e+20	2.91e+20
−17.75	2.99e+20	2.98e+20
−20.00	3.00e+20	2.81e+20
−21.75	2.99e+20	2.99e+20
−24.00	2.86e+20	3.13e+20
−25.75	3.01e+20	3.01e+20
−26.25	2.98e+20	3.02e+20
−28.00	2.87e+20	2.94e+20
−28.25	3.00e+20	3.03e+20
−28.50	2.99e+20	3.03e+20

positions and the corresponding exposure at which the MC is generated. This MC production assumes that the downstream tracker is ND-GAr rather than the TMS, however the functionality of the PRISM analysis presented in this thesis is not significantly affected by the choice of downstream tracker. Currently, the PRISM analysis only uses the downstream tracker to measure the sign and momenta of muons exiting ND-LAr and is not used as an additional source of neutrino events. The final state particles are propagated through ND-LAr and ND-GAr by GEANT4.

3.6.4 Near Detector Parameterised Reconstruction

The ND parameterised reconstruction attempts to approximate a realistic event reconstruction in the DUNE ND using true neutrino event information provided by the GENIE and GEANT4 ND simulation. Neutrino events are only considered to be reconstructed if the true interaction vertex is within the fiducial volume of ND-LAr. The PRISM analysis defines the fiducial volume to be 150 cm from the left and right faces and the downstream edge and 50 cm from the upstream edge and the top and bottom of the volume of liquid argon. The $4\,\text{m} \times 2\,\text{m} \times 3\,\text{m}$ fiducial volume is smaller than the volume used for the analysis presented in Reference [1] and this is done to improve the acceptance of ND neutrino events as the detector moves off-axis.

The parameterised reconstructed energy of each hadron is the sum of the true energy deposits of the hadron and all of its progeny in ND-LAr. It is assumed that there are no detection thresholds for the hadrons and all the hadron energy deposits are accurately associated with the respective particle. A hadronic veto region is defined to be the outer 30 cm of the active volume of ND-LAr and events that deposit more than 30 MeV of visible hadronic energy within this veto region are assumed to be poorly reconstructed due to energy being deposited outside the detector.

Muons with energies of less than approximately 1 GeV are typically contained within ND-LAr and are reconstructed by the range of the true muon track. Contained muons reconstructed by range have their kinetic energy smeared by a Gaussian function that quantifies the resolution of ND-LAr. Higher energy muons may exit the downstream edge of ND-LAr and enter ND-GAr, where the magnetic field allows the muon energy to be reconstructed by the curvature of the track and the sign of the muon determined. The momentum of muons measured in the downstream tracker is smeared by the assumed resolution of ND-GAr. If the muon in an event exits ND-LAr out of the sides or fails to propagate into ND-GAr, then that event is not reconstructed. Figure 3.17 shows the GEANT4 display of ND-LAr and ND-GAr in which neutrino events are reconstructed.

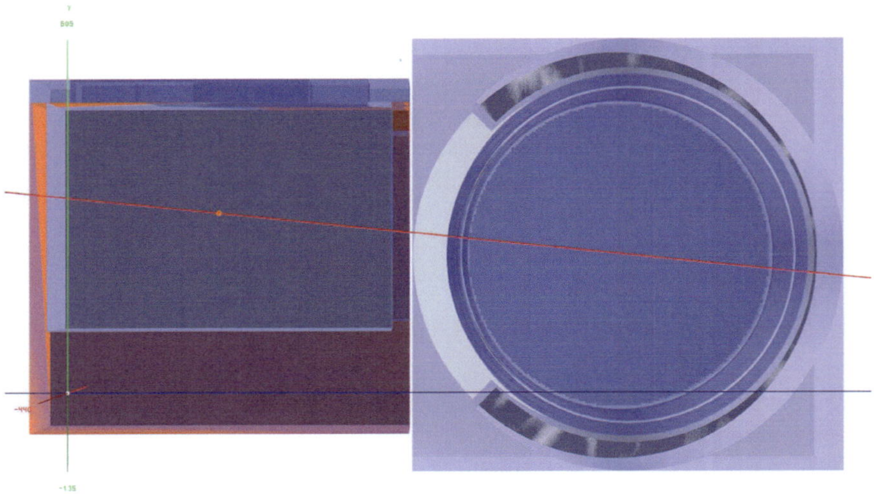

Fig. 3.17 Side view of the simulated ND complex. On the left is ND-LAr and on the right is ND-GAr. The neutrino beam axis is shown in red passing through the centre of both detectors [1]

References

1. DUNE Collaboration, B. Abi, et al., Deep Underground Neutrino Experiment (DUNE), Far Detector Technical Design Report, Volume II: DUNE Physics, Feb. 2020. arXiv: 2002.03005 [hep-ex]
2. C. Rubbia, *The liquid-argon time projection chamber: a new concept for neutrino detectors* (Technical Report, CERN, Geneva, 1977)
3. C. Rubbia et al., Underground operation of the icarus t600 lar-tpc: first results. J. Instrum. **6**(07), P07011 (2011). (Jul.)
4. MicroBooNE Collaboration, R. Acciarri, et al., "Design and Construction of the MicroBooNE Detector," JINST, vol. 12, no. 02, P02017, 2017. arXiv: 1612.05824 [physics.ins-det]
5. S. Sacerdoti, "A LArTPC with Vertical Drift for the DUNE Far Detector," PoS, vol. NuFact2021, p. 173, 2022
6. J. Asaadi et al., "First Demonstration of a Pixelated Charge Readout for Single-Phase Liquid Argon Time Projection Chambers," Instruments, vol. 4, no. 1, p. 9, 2020. arXiv: 1801.08884 [physics.ins-det]
7. A. Szelc, Developing lar scintillation light collection ideas in the short baseline neutrino detector. J. Instrum. **11**(02), C02018 (2016). (Feb.)
8. M. Soderberg, "MicroBooNE: A New Liquid Argon Time Projection Chamber Experiment," AIP Conf. Proc., vol. 1189, no. 1, F. Sanchez, M. Sorel, and L. Alvarez-Ruso, Eds., pp. 83–87, 2009. arXiv: 0910.3497 [physics.ins-det]
9. MicroBooNE Collaboration, Microboone at work: Event displays, https://microboone-exp.fnal.gov/public/approved_plots/Event_Displays.html, Accessed: 2023-04-20
10. Particle Data Group, P. A. Zyla, et al., "Review of Particle Physics," Progress of Theoretical and Experimental Physics, vol. 2020, no. 8, Aug. 2020, 083C01. eprint: https://academic.oup.com/ptep/article-pdf/2020/8/083C01/34673722/ptaa104.pdf
11. C. Anderson et al., Analysis of a large sample of neutrino-induced muons with the argoneut detector. J. Instrum. **7**(10), P10020 (2012). (Oct.)
12. E. Aprile, W. Ku, J. Park, Delta electron production and the ultimate energy resolution of liquid argon ionization detectors. IEEE Trans. Nucl. Sci. **35**(1), 37–41 (1988)

13. DUNE Collaboration, A. A. Abud, et al., “Design, construction and operation of the ProtoDUNE-SP Liquid Argon TPC,” JINST, vol. 17, no. 01, P01005, 2022. arXiv: 2108.01902 [physics.ins-det]
14. R. Acciarri et al., Oxygen contamination in liquid argon: Combined effects on ionization electron charge and scintillation light. J. Instrum. **5**(05), P05003 (2010). (May)
15. A. Zhang et al., “Modeling impurity concentrations in liquid argon detectors,” Nuclear Instruments and Methods in Physics Research Section A: Accelerators, Spectrometers, Detectors and Associated Equipment, vol. 1010, p. 165 491, 2021
16. MicroBooNE Collaboration, P. Abratenko, et al., “Measurement of the longitudinal diffusion of ionization electrons in the microboone detector,” Journal of Instrumentation, vol. 16, no. 09, P09025, Sep. 2021
17. A. Lister, M. Stancari, Investigations on a fuzzy process: Effect of diffusion on calibration and particle identification in liquid argon time projection chambers. J. Instrum. **17**(07), P07016 (2022). (Jul.)
18. MicroBooNE Collaboration, P. Abratenko, et al., “Measurement of space charge effects in the microboone lartpc using cosmic muons,” Journal of Instrumentation, vol. 15, no. 12, P12037, Dec. 2020
19. A. Hitachi et al., “Effect of ionization density on the time dependence of luminescence from liquid argon and xenon,” Phys. Rev. B, vol. 27, pp. 5279–5285, 9 May 1983
20. DUNE Collaboration, P. Adamson, et al., “Long-Baseline Neutrino Facility (LBNF)/DUNE Conceptual Design Report: Annex 3A_opt.,” 2017
21. M. Ball et al., “The PIP-II Conceptual Design Report,” V. Lebedev, Ed., Mar. 2017
22. L. Fields and A. Weber, “Beam optimization task force final report,” Tech. Rep. DUNE-doc-2901-v5, 2017, https://docs.dunescience.org/cgi-bin/sso/RetrieveFile?docid=2901&filename=BOTF_PostReview.pdf&version=5
23. DUNE Collaboration, J. Strait, et al., “Long-Baseline Neutrino Facility (LBNF) and Deep Underground Neutrino Experiment (DUNE): Conceptual Design Report, Volume 3: Long-Baseline Neutrino Facility for DUNE June 24, 2015,” Jan. 2016. arXiv: 1601.05823 [physics.ins-det]
24. DUNE Collaboration, R. Acciarri, et al., “Long-Baseline Neutrino Facility (LBNF) and Deep Underground Neutrino Experiment (DUNE): Conceptual Design Report, Volume 4 The DUNE Detectors at LBNF,” Jan. 2016. arXiv: 1601.02984 [physics.ins-det]
25. DUNE Collaboration, B. Abi, et al., “Volume iv. the dune far detector single-phase technology,” Journal of Instrumentation, vol. 15, no. 08, T08010, Aug. 2020
26. D. S. Akerib et al., “The LUX-ZEPLIN (LZ) Experiment,” Nucl. Instrum. Meth. A, vol. 953, p. 163 047, 2020. arXiv: 1910.09124 [physics.ins-det]
27. A. Machado, E. Segreto, Arapuca a new device for liquid argon scintillation light detection. J. Instrum. **11**(02), C02004 (2016). (Feb.)
28. A. Machado et al., The x-arapuca: An improvement of the arapuca device. J. Instrum. **13**(04), C04026 (2018). (Apr.)
29. W. Wu, Felix: The new detector interface for the atlas experiment. IEEE Trans. Nucl. Sci. **66**(7), 986–992 (2019)
30. MicroBooNE Collaboration, C. Adams, et al., “Calibration of the charge and energy loss per unit length of the MicroBooNE liquid argon time projection chamber using muons and protons,” JINST, vol. 15, no. 03, P03022, 2020. arXiv: 1907.11736 [physics.ins-det]
31. MicroBooNE Collaboration, C. Adams, et al., “A method to determine the electric field of liquid argon time projection chambers using a UV laser system and its application in MicroBooNE,” JINST, vol. 15, no. 07, P07010, 2020. arXiv: 1910.01430 [physics.ins-det]
32. DUNE Collaboration, A. A. Abud, et al., “Deep Underground Neutrino Experiment (DUNE) Near Detector Conceptual Design Report,” Instruments, vol. 5, no. 4, p. 31, 2021. arXiv: 2103.13910 [physics.ins-det]
33. R. Flight, “DUNE Near Detector Cavern plans/drawings,” Tech. Rep. DUNE-doc-18180-v11, 2020, https://docs.dunescience.org/cgi-bin/sso/RetrieveFile?docid=18180&filename=ND+Hall+Proposal+2021-01-25.pdf&version=14

34. G. Valdiviesso, "Status of the Short-Baseline Near Detector at Fermilab," PoS, vol. NuFact2021, p. 184, 2022
35. ArgonCube Collaboration, C. Amsler, et al., "Argoncube: A novel, fully-modular approach for the realization of large-mass liquid argon tpc neutrino detectors," 2015
36. D. Dwyer et al., Larpix: Demonstration of low-power 3d pixelated charge readout for liquid argon time projection chambers. J. Instrum. **13**(10), P10007 (2018). (Oct.)
37. N. Anfimov et al., Development of the light collection module for the liquid argon time projection chamber (lartpc). J. Instrum. **15**(07), C07022 (2020). (Jul.)
38. MINERvA Collaboration, J. Devan, et al., "Measurements of the inclusive neutrino and antineutrino charged current cross sections in MINERvA using the low-ν flux method," Phys. Rev. D, vol. 94, p. 112 007, 11 Dec. 2016
39. K. Aamodt et al., The ALICE experiment at the CERN LHC. JINST **3**, S08002 (2008)
40. DUNE Collaboration, A. Abed Abud, et al., "A Gaseous Argon-Based Near Detector to Enhance the Physics Capabilities of DUNE," Mar. 2022. arXiv: 2203.06281 [hep-ex]
41. M. Vicenzi, "SAND - System for on-Axis Neutrino Detection - in the DUNE Near Detector Complex," PoS, vol. NuFact2021, p. 248, 2022
42. D. Andrews et al., "Progress in the design, manufacture and testing of the kloe solenoid for the da/spl phi/ne ring at frascati," in Proceedings of the 1997 Particle Accelerator Conference (Cat. No.97CH36167), vol. 3, 1997, 3413–3415 vol.3
43. M. Adinolfi et al., The kloe electromagnetic calorimeter. Nucl. Instrum. Methods Phys. Res., Sect. A **482**(1), 364–386 (2002)
44. T2K Collaboration, K. Abe, et al., "The T2K Experiment," Nucl. Instrum. Meth. A, vol. 659, pp. 106–135, 2011. arXiv: 1106.1238 [physics.ins-det]
45. D. S. Ayres et al., "The NOvA Technical Design Report," Oct. 2007
46. S. Bhadra et al., "Letter of Intent to Construct a nuPRISM Detector in the J-PARC Neutrino Beamline," Dec. 2014. arXiv: 1412.3086 [physics.ins-det]
47. C. Andreopoulos et al., "The GENIE Neutrino Monte Carlo Generator," Nucl. Instrum. Meth. A, vol. 614, pp. 87–104, 2010. arXiv: 0905.2517 [hep-ph]
48. E. Snider and G. Petrillo, "Larsoft: Toolkit for simulation, reconstruction and analysis of liquid argon tpc neutrino detectors," Journal of Physics: Conference Series, vol. 898, no. 4, p. 042 057, Oct. 2017
49. B. Abi et al., First results on protodune-sp liquid argon time projection chamber performance from a beam test at the cern neutrino platform. J. Instrum. **15**(12), P12004 (2020). (Dec.)
50. J. S. Marshall and M. A. Thomson, "The Pandora Software Development Kit for Pattern Recognition," Eur. Phys. J. C, vol. 75, no. 9, p. 439, 2015. arXiv: 1506.05348 [physics.data-an]
51. MicroBooNE Collaboration, R. Acciarri, et al., "The Pandora multi-algorithm approach to automated pattern recognition of cosmic-ray muon and neutrino events in the MicroBooNE detector," Eur. Phys. J. C, vol. 78, no. 1, p. 82, 2018. arXiv: 1708.03135 [hep-ex]

Chapter 4
Data-Driven Far Detector Prediction

DUNE will measure the disappearance of ν_μ and $\bar{\nu}_\mu$ neutrinos and the appearance of ν_e and $\bar{\nu}_e$ neutrinos in the LBNF beam. As stated in Sect. 3.5, the PRISM analysis aims to make this measurement by building linear combinations of on and off axis ND data that accurately predict the FD event rate. This chapter will describe in detail how data-driven predictions of the FD ν_μ, $\bar{\nu}_\mu$, ν_e and $\bar{\nu}_e$ event rates are produced using the PRISM technique. The process of producing a PRISM prediction of the FD event rate is summarised in Fig. 4.1 and each step in the diagram will be explained in detail.

4.1 Data Selection

Although all the data used in this analysis is MC, a distinction is made between MC that is labelled as "data" or "MC" in order to simulate a realistic PRISM analysis. Here, the data represents events that would be measured in the detectors and MC refers to where simulated data would be used, for example in background predictions. The same analysis variables and selection criteria are used for data and MC samples set out in this section.

4.1.1 Analysis Variables

The neutrino oscillation probability depends on the energy of the neutrino. However, it is not possible for experiments to directly measure the energy of the neutrino. Instead, the neutrino energy is inferred from the observable final state particles in a neutrino interaction. In order to maximise the sensitivity to the oscillation parameters,

C. M. Hasnip, *DUNE-PRISM—A New Method to Measure Neutrino Oscillations*, Springer Theses, https://doi.org/10.1007/978-3-032-07251-1_4

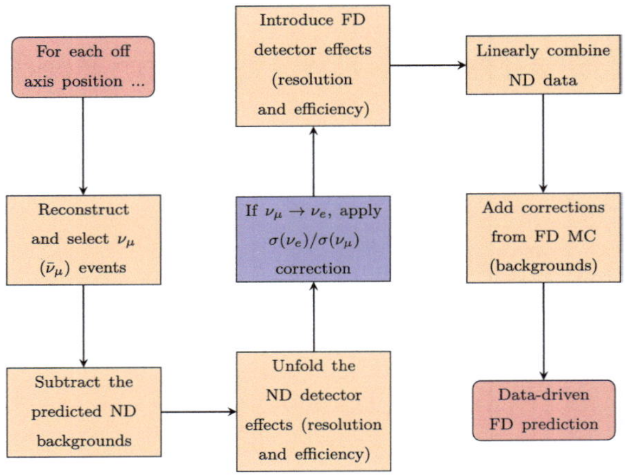

Fig. 4.1 Flow diagram summarising the process of producing a PRISM prediction of the ν_μ, $\bar{\nu}_\mu$, ν_e and $\bar{\nu}_e$ FD event rate. The start of the process is the top left box. The blue box is only performed for ν_e and $\bar{\nu}_e$ appearance predictions

the variable with which the neutrino oscillation analysis is conducted should be closely related to the neutrino energy and be constructed from quantities observable in the detector. The neutrino interaction model is then relied upon to relate the final state particles to the initial neutrino energy. The DUNE FD uses the variable "reconstructed neutrino energy", which sums the reconstructed energy from the final state lepton in a CC interaction $E^{rec.}_{lep.}$ and the hadronic activity emanating from the target nucleus $E^{rec.}_{had.}$:

$$E^{rec.}_{FD} = E^{rec.}_{lep.} + E^{rec.}_{had.}. \tag{4.1}$$

In an oscillation analysis it is desirable to use the same analysis variable in the near and far detectors. However, as discussed in Sect. 3.6, a limitation of the current analysis is that full event reconstruction is only available in the FD MC. Consequently, a different variable is used in the DUNE ND called the "visible reconstructed neutrino energy". Like the FD variable, the visible reconstructed energy sums the reconstructed lepton and hadronic energy. Neutral hadrons that are difficult to reconstruct, such as neutrons, are ignored in this variable. The form of the ND analysis variable is

$$E^{rec.}_{ND} = E^{rec.}_{\mu} + E^{rec.}_{p} + E^{rec.}_{\pi^\pm} + E^{rec.}_{\pi^0} + E^{rec.}_{other}, \tag{4.2}$$

where $E^{rec.}_{\mu}$, $E^{rec.}_{p}$, $E^{rec.}_{\pi^\pm}$, $E^{rec.}_{\pi^0}$ and $E^{rec.}_{other}$ are the reconstructed energies of the muon, proton, charged pion, neutral pion and other visible hadronic energy respectively. Reconstructed energy in the case of the $E^{rec.}_{ND}$ analysis variable refers to the ND parameterised reconstruction described in Sect. 3.6.4.

An additional variable called visible true energy ($E^{true}_{vis.}$) is defined, which is the true energy analogy to $E^{rec.}_{ND}$ and is given by

$$E^{true}_{vis.} = E^{true}_{\mu} + T^{true}_{p} + T^{true}_{\pi^\pm} + T^{true}_{\pi^0} + T^{true}_{other}, \tag{4.3}$$

where E_{μ}^{true} is the true total energy of the muon and T_{p}^{true}, $T_{\pi^{\pm}}^{true}$, $T_{\pi^{0}}^{true}$ and T_{other}^{true} are the true kinetic energies of the proton, charged pions, neutral pion and other charged hadrons respectively. Reconstructed and true *visible* energy variables are chosen for the ND in order to work with quantities that are observable in the detector, thereby limiting any dependence on the neutrino interaction model that could be introduced by the correction for ND resolution and acceptance effects that will be discussed in Sect. 4.3.2. However, it should be noted that $E_{FD}^{rec.}$ does include model dependent corrections for unobserved particle energy deposits in an event. These corrections improve the resolution of $E_{FD}^{rec.}$ to the true neutrino energy, thereby increasing the precision of the oscillation parameter measurement. The reconstructed neutrino energy defined in Eq. 4.1 is the same variable used for the DUNE Technical Design Report (TDR) oscillation analysis [1]. This analysis keeps the same $E_{FD}^{rec.}$ variable as the TDR to enable direct comparisons between PRISM and existing DUNE oscillation sensitivity studies.

4.1.2 Far Detector CVN Selection

All neutrino events are required to originate within the fiducial volume of the FD module. The disappearance analysis selects ν_{μ} and $\bar{\nu}_{\mu}$ events and the appearance analysis selects ν_{e} and $\bar{\nu}_{e}$ events in the FD. When the beam is running in FHC mode the beam is enriched in neutrinos and in RHC mode the beam is enriched with antineutrinos. The FD selection relies on a Convolutional Visual Network (CVN) developed by the DUNE Collaboration to select the primary lepton in a CC neutrino interaction based on images produced by the FD reconstruction. The CVN is an image recognition machine learning technique well-suited to the event displays produced by LArTPCs and more detail on the CVN design can be found in Ref. [2]. Training was performed on 500×500 pixel images of neutrino interactions from each of the three wire planes in the FD. The wire planes provided images of wire number versus time, where each pixel in the image corresponded to the total charge deposited in a reconstructed hit. No higher-level reconstructed objects were required for the CVN to learn how to identify the primary charged lepton in a CC interaction. The CVN outputs a score for each event that quantifies the probability that the neutrino is a ν_{μ} or ν_{e} in FHC mode and a $\bar{\nu}_{\mu}$ or $\bar{\nu}_{e}$ in RHC mode. The output from the CVN for muon and electron neutrino selection in the FD is shown in Fig. 4.2.

If an event is selected by the CVN to be either a ν_{μ} or $\bar{\nu}_{\mu}$ then the reconstructed energy of the neutrino is the sum of the reconstructed energy of the longest track and the reconstructed hadronic energy, as stated in Sect. 3.6.2 and Eq. 4.1. The cut applied to select ν_{μ} and $\bar{\nu}_{\mu}$ is a CVN score greater than 0.5 in both FHC and RHC mode, as can be seen in Fig. 4.2a. If the CVN identifies the event to be an electron neutrino the reconstructed energy is the sum of the highest energy electromagnetic shower and the reconstructed hadronic energy. The cut applied for the ν_{e} and $\bar{\nu}_{e}$ event selection is an electron neutrino CVN score greater than 0.85, as shown in Fig. 4.2b. Figure 4.3 shows the FD event rates for the four signal channels selected by the CVN,

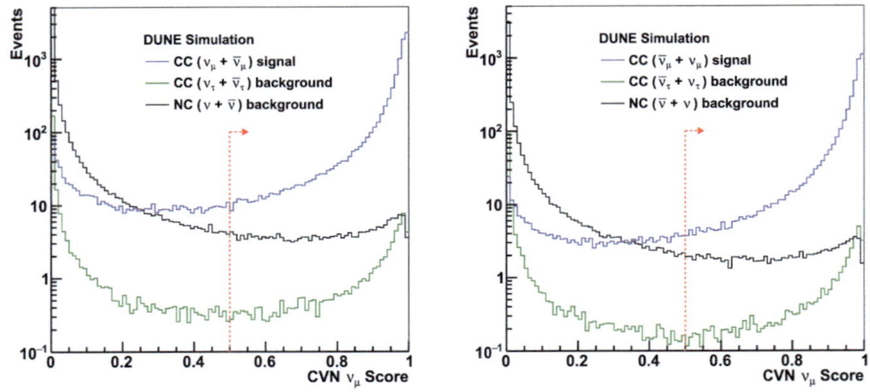

(a) CVN ν_μ and $\bar{\nu}_\mu$ selection.

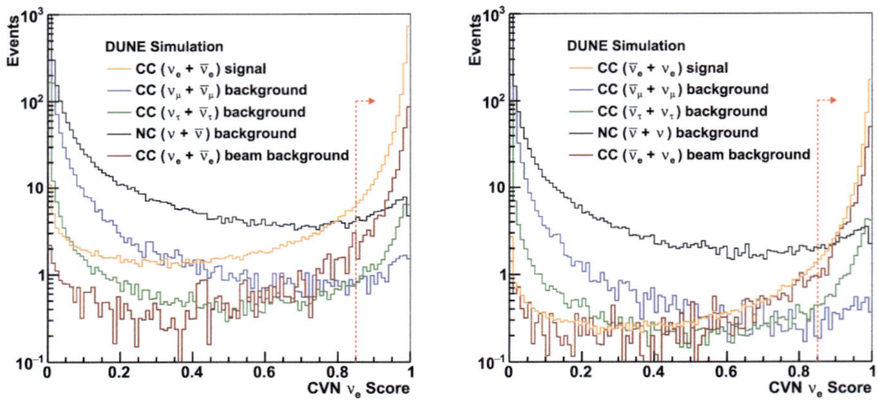

(b) CVN ν_e and $\bar{\nu}_e$ selection.

Fig. 4.2 CVN scores for FD selection of ν_μ ($\bar{\nu}_\mu$) and ν_e ($\bar{\nu}_e$) in FHC (RHC) mode. Figures from Ref. [2]. The red line shows the position of the cut applied on the CVN score of each event

where the total selected event rate has been separated into individual contributions from the true signal and the different background types.

4.1.2.1 Far Detector Backgrounds

Whilst the CVN is a highly effective selector of signal events, backgrounds do pass the selection cuts. The size of the backgrounds are predicted from the FD MC data and there are different sources of background events for muon neutrino disappearance and electron neutrino appearance measurements.

The expected FD background event rates for the muon neutrino disappearance measurement are shown in Fig. 4.3a and b. Neutral current backgrounds feature

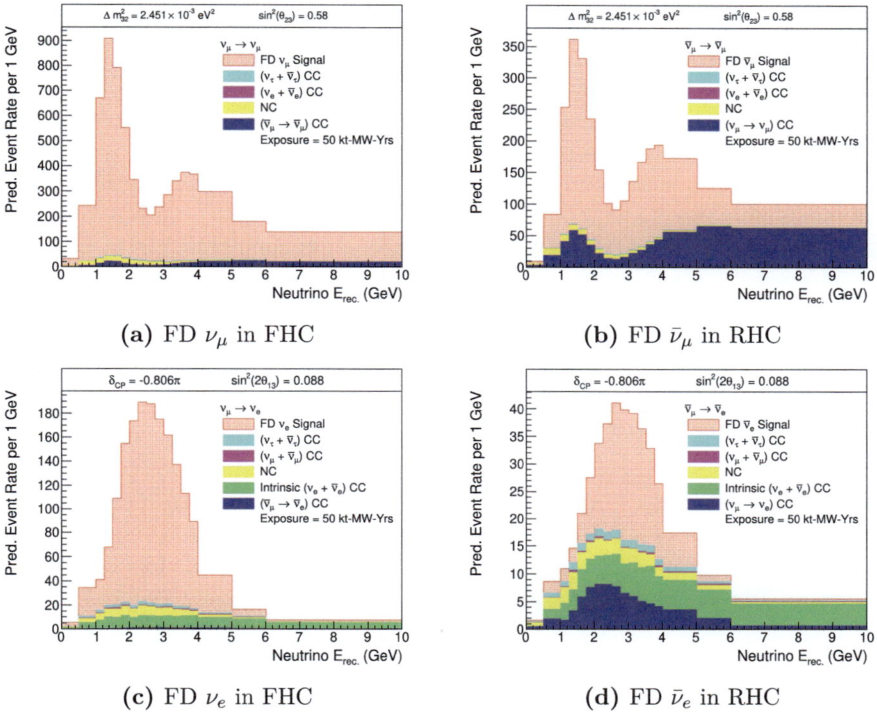

(a) FD ν_μ in FHC

(b) FD $\bar{\nu}_\mu$ in RHC

(c) FD ν_e in FHC

(d) FD $\bar{\nu}_e$ in RHC

Fig. 4.3 Event rate spectra for selected ν_μ, $\bar{\nu}_\mu$, ν_e and $\bar{\nu}_e$ events in the FD. The stacked histograms show the individual signal and background components of the total selected event rate for each signal channel. The event rates assume the NuFIT 4.0 oscillation parameters [3]

prominently at the lower energies. This is because charged pions produced in NC interactions can produce long tracks that are similar to short-range, low-energy muon tracks. The FD is unable to measure the charge of the muon and so cannot distinguish between between ν_μ and $\bar{\nu}_\mu$ on an event-by-event basis. Therefore, there is a high probability that $\bar{\nu}_\mu$ and ν_μ beam contamination in FHC and RHC mode respectively are selected as signal events. This is called "wrong-sign" background since a neutrino is selected that produces a lepton with the opposite charge to the lepton expected in a signal event. Figure 4.3b demonstrates that backgrounds form a much larger component of the total selected event rate in RHC mode than FHC mode. This is due to the large wrong-sign background in the $\bar{\nu}_\mu$ selection in RHC mode. There is a sizeable amount of background from CC ν_τ and $\bar{\nu}_\tau$ and this is due to the tau lepton decaying rapidly into a muon. Finally, there is a small contribution from electron neutrinos being misidentified as muons. The background due to muon neutrino appearance in the beam due to $\nu_e \to \nu_\mu$ oscillations is negligible.

The ν_e and $\bar{\nu}_e$ background event rates are shown in Fig. 4.3c and d respectively. Background events in the electron neutrino selection can be induced by misidentifying photons. Photons are produced in the decay of a π^0 emanating from a NC interaction. There is no way of distinguishing the sign of an electron produced by a CC interaction in a LArTPC. It is therefore likely that wrong-sign electron neutrino backgrounds will be selected as signal. The wrong-sign background is a significant component of the background rate in RHC mode, as can be seen in Fig. 4.3d. The signal events are electron neutrinos that have appeared in beam after oscillations. Electron neutrinos intrinsic to the neutrino beam cannot be distinguished from signal events and so form a large component of the background. Tau neutrinos can decay into an electron neutrino, leading to a small background contribution from ν_τ and $\bar{\nu}_\tau$ CC interactions. Finally, a small number of muon neutrinos may be misidentified as electron neutrino events.

4.1.3 Near Detector Selection

Events in the ND are selected as ν_μ ($\bar{\nu}_\mu$) by requiring a μ^- (μ^+) that originates in the fiducial volume of ND-LAr and is either fully contained in ND-LAr or enters the ND-GAr downstream. The muon is identified by searching for the longest track in the event. The longest track must be at least 1 m and not deposit more than 3 MeV of energy per centimetre. These requirements suppress the number of charged pions from NC events that are misidentified as muons. Events with identified muons below 400 MeV have significant charged pion backgrounds and are therefore removed. Additionally, the hadronic energy is required to be well contained in the volume of ND-LAr by having no more than 30 MeV of hadronic energy deposited in the ND-LAr veto region. The charge of the muon is assumed to be perfectly selected for events where the muon propagates to ND-GAr. Therefore, for the events where the muon enters ND-GAr only events with a μ^- (μ^+) are selected in FHC (RHC) mode. The wrong-sign background contribution in FHC mode is very small, so all muon-like events stopping ND-LAr are assumed to be μ^- in FHC mode. A stopping muon contained in ND-LAr will emit a Michel electron. There is a 75% probability that a μ^- originating from the wrong-sign background in RHC mode will capture on an argon nucleus instead of stopping and decaying to an electron. Therefore, a Michel electron is required in the final state of contained ND-LAr RHC events, resulting in the contained RHC wrong-sign background being suppressed by factor of four.

The event rates of selected ν_μ and $\bar{\nu}_\mu$ events in the ND are shown in Fig. 4.4a and b respectively. Data is taken with ND-LAr and TMS/ND-GAr positioned at discrete stops, measuring neutrino events at one stop for a period of time before moving to the next position. As can be seen in Fig. 4.4, all of the data taken across each position is combined into a continuous distribution of event rates that is a function of both off-axis position and $E_{ND}^{rec.}$. Each bin in off-axis position is 50 cm-wide, but the ND-LAr fiducial volume is 4 m-wide. Hence, neutrino event rates are measured for eight off-axis position bins whilst stationary at a given stop, allowing there to be many more off-axis position bins that there are stops to physically move the ND to.

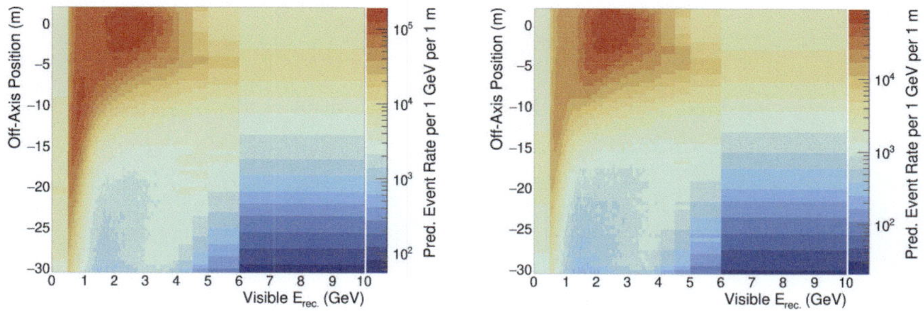

(a) ND ν_μ in FHC

(b) ND $\bar{\nu}_\mu$ in RHC

Fig. 4.4 Event rate spectra for selected ν_μ and $\bar{\nu}_\mu$ events in the ND. The ND event rate spectra in Fig. 4.4a and b are a function of visible $E_{ND}^{rec.}$ and off-axis position

4.1.3.1 Near Detector Backgrounds

Background events may occur in the ND that imitate the signal and pass the ND selection cuts. At low energies, charged pions from NC events can produce the longest track in an event and be identified as the muon in a CC muon neutrino interaction. The muon sign selection of ND-GAr suppresses the wrong-sign background at high neutrino energies. Consequently, the wrong-sign background is concentrated in the events where the longest track is contained in ND-LAr. Events where an electron is misidentified as a muon are a negligible contribution to the ND background. Figure 4.5 shows the event rates for the NC and wrong-sign backgrounds in the ND for the two horn polarities.

4.1.4 Additional Horn Current Sample

The magnitude of the current flowing through the LBNF focusing horn determines the strength of the focusing of charged pions. This analysis assumes a nominal horn current of 293 kA. However, the PRISM analysis can be improved by taking a small sample of ND data on-axis with the slightly lower horn current of 280 kA. The lower horn current focuses fewer pions that decay to high-energy on-axis neutrinos. This results in a reduction in the on-axis flux in the 3–5 GeV region, as can be seen in Fig. 4.6. The 280 kA horn current sample improves the performance of the PRISM linear combination in the 3–5 GeV region by providing an additional contribution to the linear combination at high neutrino energies. This is necessary because the highest energy neutrino flux in the nominal horn current sample is the on-axis flux, which peaks at 2.5 GeV. It is therefore challenging to build a linear combination prediction that agrees well with the measured the FD event rate at energies above

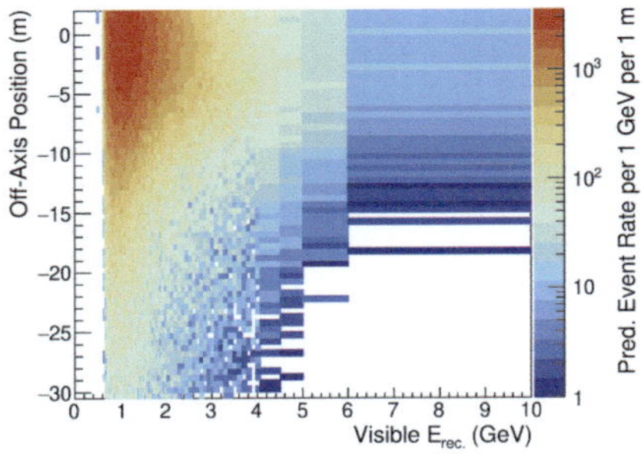

(a) NC FHC

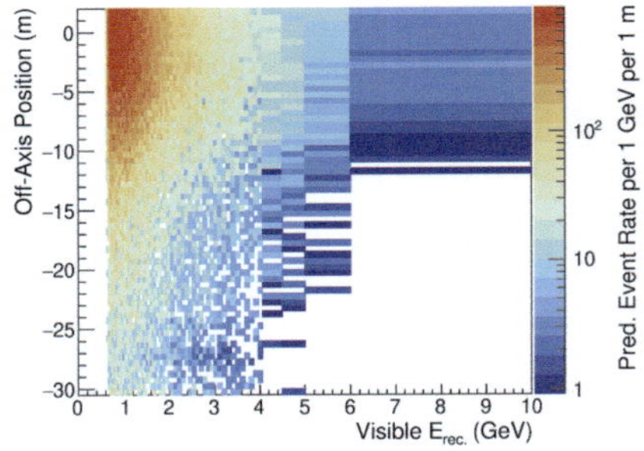

(b) NC RHC

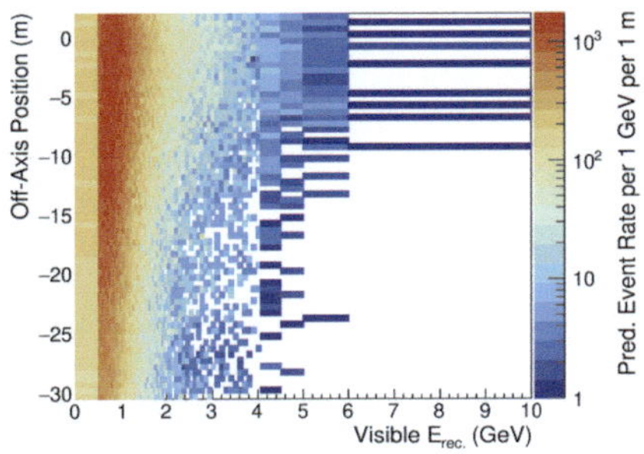

(c) Wrong-sign FHC.

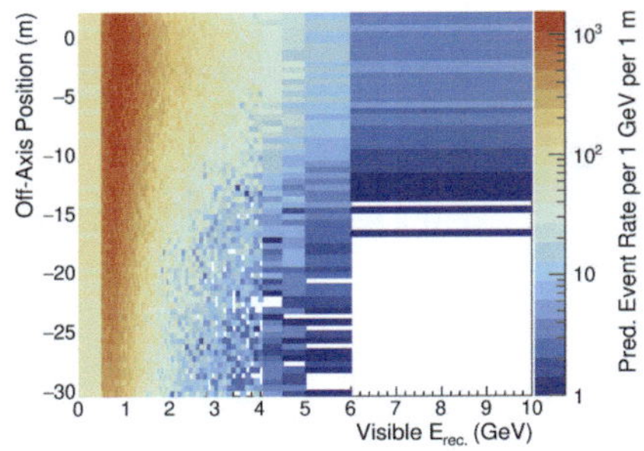

(d) Wrong-sign RHC.

Fig. 4.5 Neutral current and wrong-sign ND background event rate in the in FHC and RHC mode

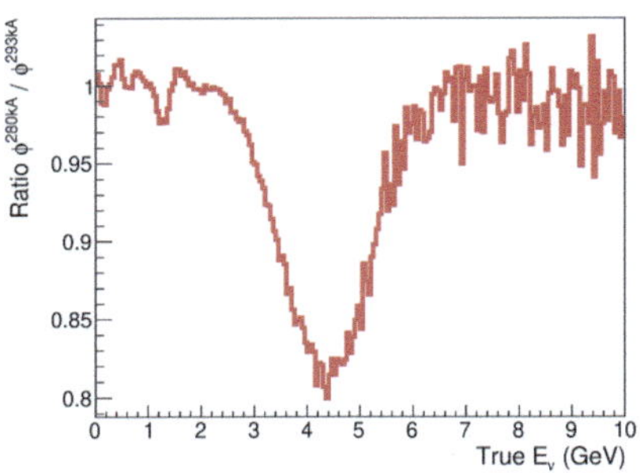

(a) FHC ν_μ ratio.

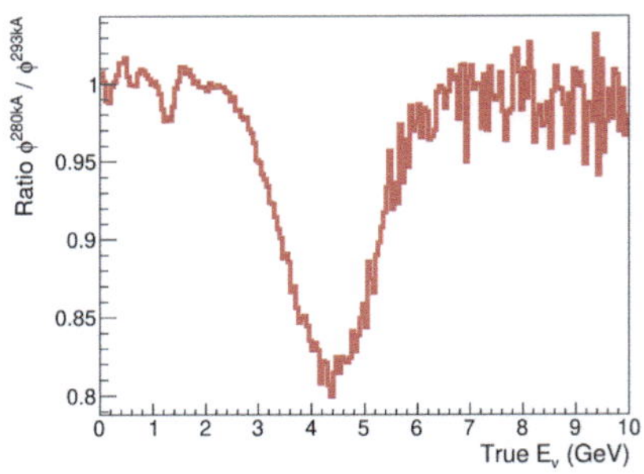

(b) RHC $\bar{\nu}_\mu$ ratio.

Fig. 4.6 Ratios of the 280 kA to the 293 kA LBNF flux at the ND for ν_μ and $\bar{\nu}_\mu$ in FHC and RHC mode respectively

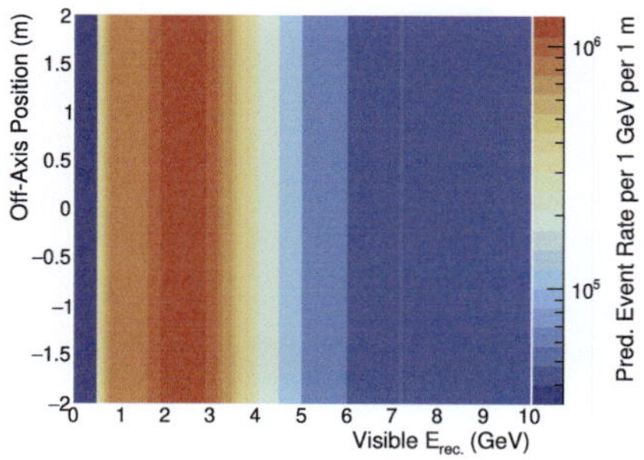

(a) FHC ν_μ

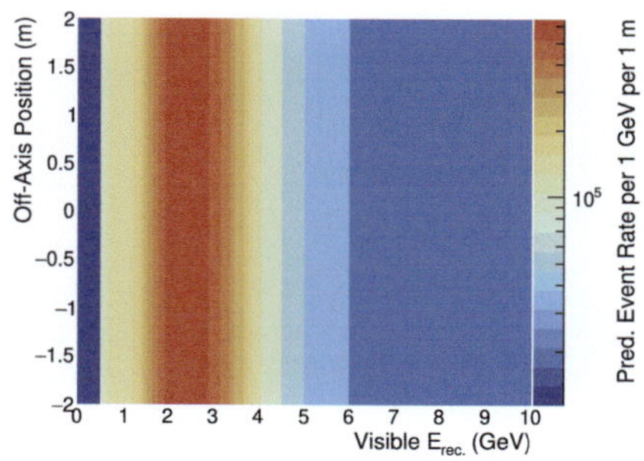

(b) RHC $\bar{\nu}_\mu$

Fig. 4.7 ND event rates on-axis for the 280 kA horn current sample

3 GeV without the additional lower horn current sample. Further details on the additional horn current sample can be found in Sect. 4.5.1 of Ref. [4].

To produce the additional 280 kA horn current ND sample true ν_μ ($\bar{\nu_\mu}$) CC on-axis events in FHC (RHC) mode are weighted by the ratio shown in Fig. 4.6a (4.6b). Exactly the same event selection criteria is applied to this weighted sample of on-axis data as for the nominal horn current data. The ND on-axis event rates for the 280 kA sample are shown in Fig. 4.7. Notice the y-axes in Fig. 4.7 have only one bin. The LBNF beam would likely only run at an alternative horn current for a short period of time, so having one y-axis bin suppresses statistical fluctuations for a short exposure of the ND to the 280 kA flux.

4.2 Near Detector Run-Plan

The amount of time ND-LAr and ND-GAr spend in each off-axis position is called the run-plan. So far, the ND event rates shown have assumed the ND spends an equal amount of time in each off-axis position. However, this is not necessarily the optimum run-plan for PRISM. A simple optimisation was performed that found it beneficial to keep the ND on-axis for the majority of each yearly run-plan and to have a smaller exposure in each off-axis position [4]. It is also the case that the standard DUNE oscillation analysis requires at least half the ND data to be taken in the on-axis position. Only a small amount of exposure is needed for the 280 kA on-axis sample. It is assumed that there will be 28 weeks of beam exposure each year and Fig. 4.8a and b show the chosen ND run-plan in terms of the number of weeks at each position and the corresponding weight respectively. The same run-plan is assumed for FHC and RHC mode. The ND data in each off-axis position is weighted according to this run-plan to account for the expected exposure at each off-axis position, as shown in Fig. 4.8c and d.

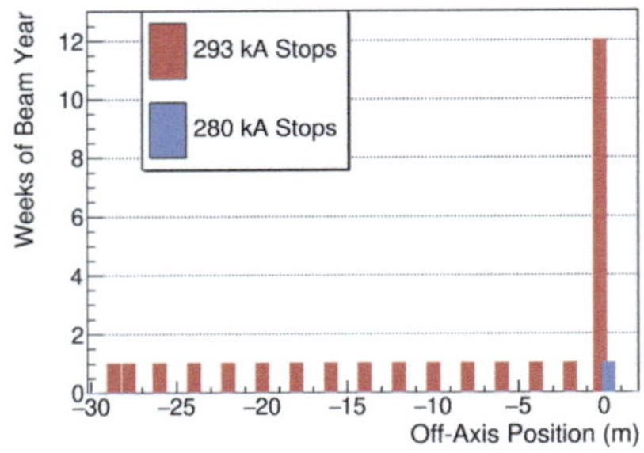

(a) ND Run-Plan Weeks

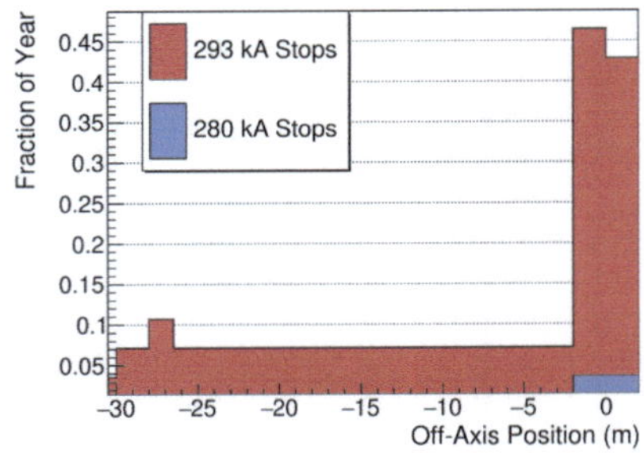

(b) ND Run-Plan Weights

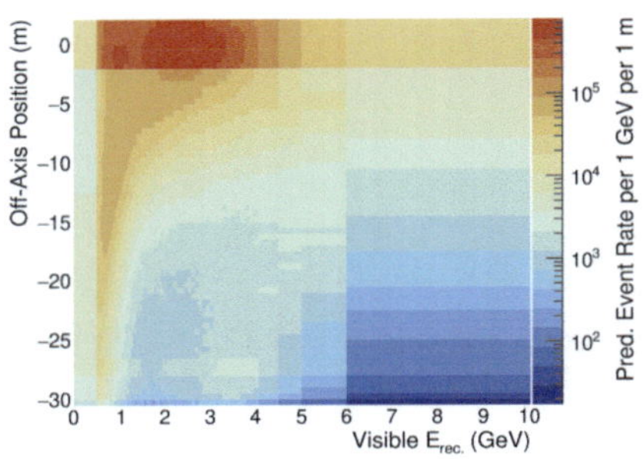

(c) FHC ν_μ

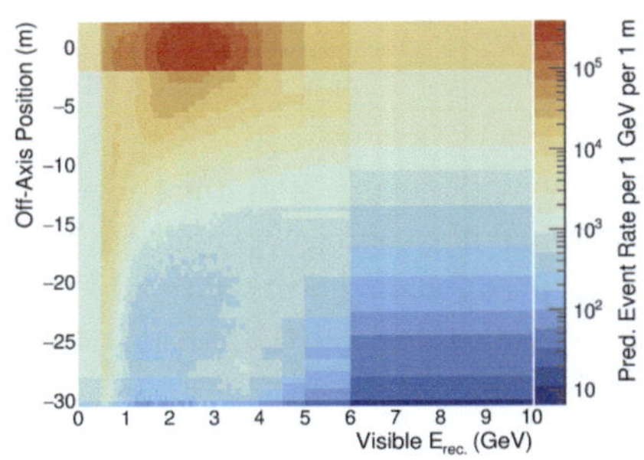

(d) RHC $\bar{\nu}_\mu$

Fig. 4.8 ND run-plan and ND event rates in FHC and RHC mode weighted for the chosen run-plan. The ND run plan is given in the number of weeks at each position in the assumed 28-week beam year (**a**) and the corresponding weights for each off-axis position (**b**). The small on-axis ND sample taken at the lower 280 kA horn current is also included in the run plan

4.3 Correcting for Detector Effects

In a PRISM analysis data will be taken at the ND for the given run-plan, such as the one shown in Fig. 4.8b. For the rest of the analysis procedure the run-plan weighted MC shown in Fig. 4.8c and d is regarded as the "data" expected to be observed by the ND. The aim of the PRISM analysis is to linearly combine this ND data to produce a data-driven prediction of the FD event rate. However, this cannot be done with raw ND data because the near and far detectors are different. Firstly, the background event rates do not extrapolate to the FD in the same way as the signal due to neutrino oscillations. The ND and FD are also very different sizes, so whilst practically all neutrino events occurring in the FD fiducial volume are well-contained this is far from the case in the ND. Although both detector complexes used liquid argon technology, the detector designs are not the same, which could lead to different resolutions. A method for correcting for the differences between the ND and FD has been developed for this thesis.

4.3.1 Near Detector Background Subtraction

The first step is to subtract the ND background event rates away from each slice of ND off-axis data. A "slice" refers to a single histogram bin in off-axis position at the ND. The background event rates at each off axis position are predicted by the MC simulation. Subtracting the ND backgrounds leaves an approximately pure sample of ν_μ ($\bar{\nu}_\mu$) signal events in each slice of ND data in FHC (RHC) mode. Fig. 4.9 shows

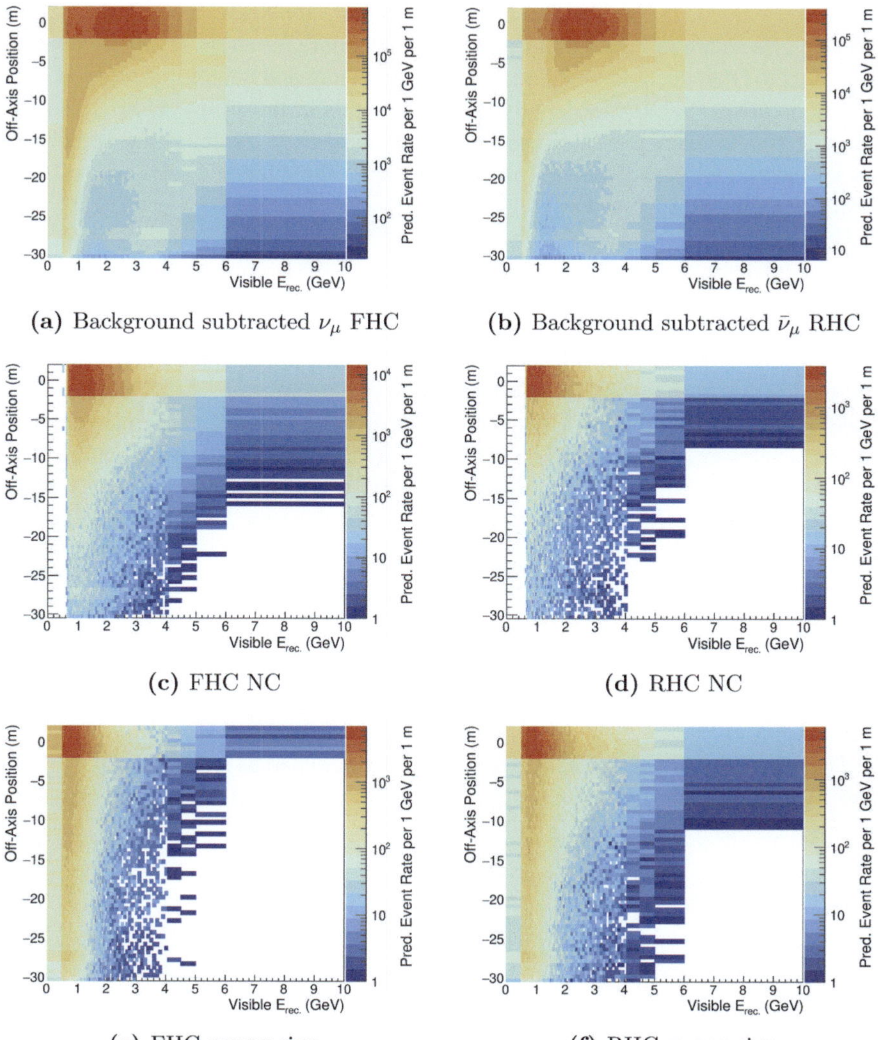

(a) Background subtracted ν_μ FHC

(b) Background subtracted $\bar{\nu}_\mu$ RHC

(c) FHC NC

(d) RHC NC

(e) FHC wrong-sign

(f) RHC wrong-sign

Fig. 4.9 Background subtracted ND signal for the chosen run-plan and both horn polarities. Run-plan weighted predicted background events that were subtracted from the ND data also shown the predicted ND backgrounds that are subtracted from the selected ND data and the resulting background subtracted ND data for the 293 kA nominal horn current sample. The same process of background subtraction is performed for the 280 kA on-axis sample.

4.3.2 Removing Near Detector Effects

The ND effects are defined by the selection efficiency and detector resolution. These detector effects are removed by an unfolding procedure based on Tikhonov regularisation [5, 6] that accounts for both the ND efficiency and detector resolution. The selection cuts at the ND are largely acceptance cuts in the simple event selection. The selection efficiency is calculated as a function of $E_{vis.}^{true}$, which is the true visible energy calculated from Eq. 4.3. The ND selection efficiency at a particular off-axis position in the jth $E_{vis.}^{true}$ bin is

$$\epsilon_j = \frac{\text{Selected signal events}}{\text{Total true signal events}}. \tag{4.4}$$

The selection efficiency at each off-axis position for the nominal horn current sample and for the two horn polarities is shown in Fig. 4.10a and b.

The true energy of the final state particles in an event is indifferent to the detector design. Therefore, each slice of ND data is unfolded using a smearing matrix $\mathbf{M}^{\mathbf{ND}}$ derived from the ND MC. The smearing matrix maps the jth bin of true visible energy E_j^{true} to the ith bin of reconstructed visible energy $E_i^{rec.}$ in the following way:

$$E_i^{rec.} = \sum_j^{n_{true}} M_{ij}^{ND} E_j^{true}, \tag{4.5}$$

where the "ND" and "vis." subscripts have been dropped from $E_{ND}^{rec.}$ and $E_{vis.}^{true}$ respectively.

The efficiency of the ND selection, which is calculated as a function of $E_{true}^{vis.}$, is also corrected for during the unfolding. This is done by normalising the smearing matrix such that

$$\sum_i^{n_{rec.}} M_{ij}^{ND} = \epsilon_j, \tag{4.6}$$

where the sum of the reconstructed energy elements for a particular E_j^{true} bin gives the efficiency ϵ_j in that E_j^{true} bin. Examples of ND smearing matrices for FHC and RHC mode normalised to the selection efficiency on-axis are shown in Fig. 4.10c and d respectively. The ND data is unfolded one off-axis slice at a time to account for the selection efficiency in the particular off-axis slice being unfolded. This is due

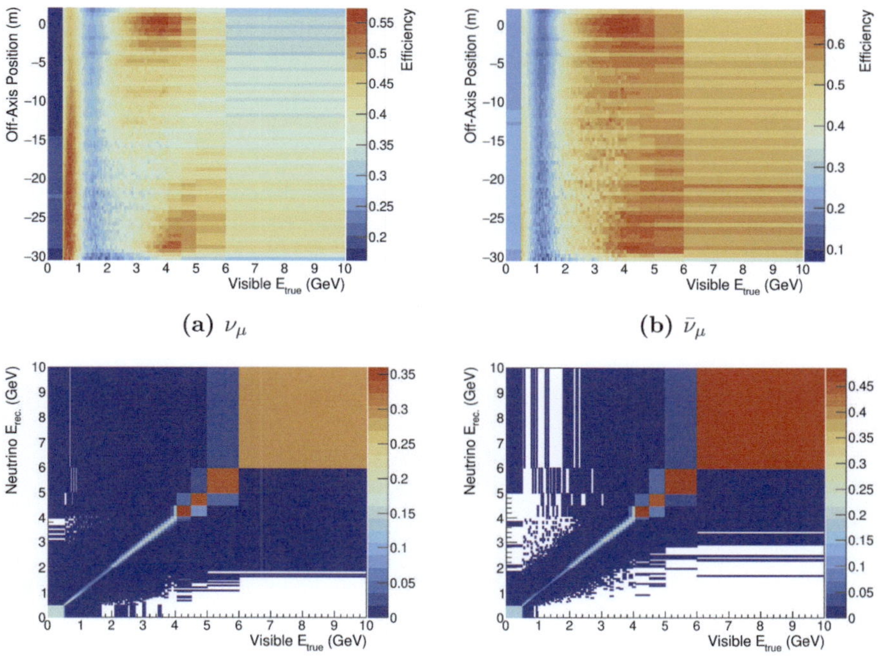

(a) ν_μ (b) $\bar{\nu}_\mu$

(c) ν_μ (d) $\bar{\nu}_\mu$

Fig. 4.10 ND selection efficiency at each off-axis position as a function of $E^{true}_{vis.}$ and ND smearing matrices normalised to the efficiency at the on-axis position

to the large variations in selection efficiency with off-axis position, as can be seen in Fig. 4.10. The resolution of the ND is assumed to be the same in each position.

The simplest way to unfold the ND effects would be to minimise the sum of squared residuals:

$$\left\| \mathbf{M^{ND}} E^{true} - E^{rec.} \right\|^2, \tag{4.7}$$

which is equivalent to solving Eq. 4.5 for E^{true} by matrix inversion. The notation $\|\cdot\|$ denotes the Euclidean norm. The unfolded solution would therefore be

$$E^{true} = \left(\mathbf{M^{ND}}\right)^{-1} E^{rec.}. \tag{4.8}$$

If followed by the FD smearing transformation described in Sect. 4.3.4, this can result in an acceptable solution. However, the matrix inversion method results in large variances in E^{true} solution that can be exacerbated by systematic uncertainty parameters that alter the relationship between E^{true} and $E^{rec.}$, such as energy scale uncertainties. Hence, to provide greater numerical stability to the unfolding procedure a regularisation condition is added to Eq. 4.7. The quantity to minimise is now

$$\left\|\mathbf{M}^{\mathbf{ND}} E^{true} - E^{rec.}\right\|^2 + \left\|\Gamma E^{true}\right\|^2, \tag{4.9}$$

where the second term penalises solutions in which the difference between adjacent elements changes rapidly. This is achieved through the regularisation matrix, Γ, that has the form

$$\Gamma = \tau_{unf.} \begin{pmatrix} 1 & -2 & 1 & 0 & \cdots & 0 & 0 & 0 \\ 0 & 1 & -2 & 1 & \cdots & 0 & 0 & 0 \\ 0 & 0 & 1 & -2 & \cdots & 0 & 0 & 0 \\ 0 & 0 & 0 & 1 & \cdots & 0 & 0 & 0 \\ \vdots & \vdots & \vdots & \vdots & \ddots & \vdots & \vdots & \vdots \\ 0 & 0 & 0 & 0 & \cdots & 1 & -2 & 1 \\ 0 & 0 & 0 & 0 & \cdots & 0 & 0 & 0 \\ 0 & 0 & 0 & 0 & \cdots & 0 & 0 & 0 \end{pmatrix}, \tag{4.10}$$

where the strength of the regularisation is determined by the parameter $\tau_{unf.}$. In other words, the choice of Γ in Eq. 4.9 suppresses rapid changes in the curvature of the solution. There is a risk in Tikhonov unfolding of introducing bias into the solution by making the regularisation condition too strong. To avoid this, $\tau_{unf.}$ is made as small as possible, whilst just being big enough to suppress the largest variations in E^{true}. From Eq. 4.9 it can be shown [7, 8] that the inverted ND smearing matrix used to unfold $E^{rec.}$ in Eq. 4.8 is replaced by the unfolding matrix

$$\mathbf{D} = \left(\left(\mathbf{M}^{\mathbf{ND}}\right)^{\mathrm{T}} \mathbf{M}^{\mathbf{ND}} + \Gamma^{\mathrm{T}}\Gamma\right)^{-1} \left(\mathbf{M}^{\mathbf{ND}}\right)^{\mathrm{T}}, \tag{4.11}$$

where the unfolded solution is now

$$E^{true} = \mathbf{D}E^{rec.}. \tag{4.12}$$

Poisson distributed statistics are assumed for each bin of ND data that is unfolded. Unfolding each off-axis bin of ND data can introduce correlations between the energy bins, which are accounted for by describing the statistical uncertainty through a covariance matrix. A diagonal covariance matrix is defined for each slice of off-axis data, $\mathbf{V}^{\mathbf{ND}}$, which is propagated through the Tikhonov unfolding procedure using the unfolding matrix defined in Eq. 4.11. The corresponding covariance matrix for the unfolded solution E^{true} at a particular off-axis position is therefore

$$\mathbf{V}^{\mathbf{unf.}} = \mathbf{D}\mathbf{V}^{\mathbf{ND}}\mathbf{D}^{\mathrm{T}}, \tag{4.13}$$

where $\mathbf{D}$ is calculated from Eq. 4.11.

4.3.3 $\sigma(\nu_e)/\sigma(\nu_\mu)$ Cross Section Ratio

The step described in this section is only required for the electron neutrino appearance analysis, as shown in the central blue box of Fig. 4.1. The PRISM analysis builds data-driven predictions of the FD event rate from linear combinations of off-axis muon neutrino measurements at the ND. This is the case for both the disappearance and appearance measurement. Therefore, an additional step is required for the PRISM ν_e ($\bar{\nu}_e$) appearance analysis that corrects for the difference between the ν_μ ($\bar{\nu}_\mu$) and ν_e ($\bar{\nu}_e$) cross sections. A future version of the PRISM analysis will include a data-driven measurement of the $\sigma(\nu_e)/\sigma(\nu_\mu)$ and $\sigma(\bar{\nu}_e)/\sigma(\bar{\nu}_\mu)$ cross section ratios. Currently, a simple MC-based correction is implemented.

The cross section ratios, $\sigma(\nu_e)/\sigma(\nu_\mu)$ and $\sigma(\bar{\nu}_e)/\sigma(\bar{\nu}_\mu)$, are calculated as a function of $E^{vis.}_{true}$ by dividing an electron neutrino event rate sample by a muon neutrino event rate sample with an identical flux. The event rates are not dependent on the detector model, since they are a function of true MC quantities and all true ν_μ and ν_e CC signal events are selected. The event rates are therefore simply $flux \times cross\ section$ and the identical ν_μ and ν_e fluxes cancel, leaving the cross section ratio. The theoretical prediction of the cross section ratios [9] is, of course, a function of neutrino energy, not $E^{vis.}_{true}$. The ratios implemented in this analysis and shown in Fig. 4.11 can therefore be regarded as empirical corrections based on the MC. Larger ν_e ($\bar{\nu}_e$) cross sections at low neutrino energies are accounted for by the increased phase space in the ν_e ($\bar{\nu}_e$) cross section calculation due to the small electron mass [10]. Each slice of off-axis ND data, which has been unfolded to the $E^{vis.}_{true}$ variable, is weighted by the cross section ratios shown in Fig. 4.11. In the ν_e appearance measurement the $\sigma(\nu_e)/\sigma(\nu_\mu)$ ratio is applied (Fig. 4.11a) and in the $\bar{\nu}_e$ appearance measurement the $\sigma(\bar{\nu}_e)/\sigma(\bar{\nu}_\mu)$ ratio is applied (Fig. 4.11b).

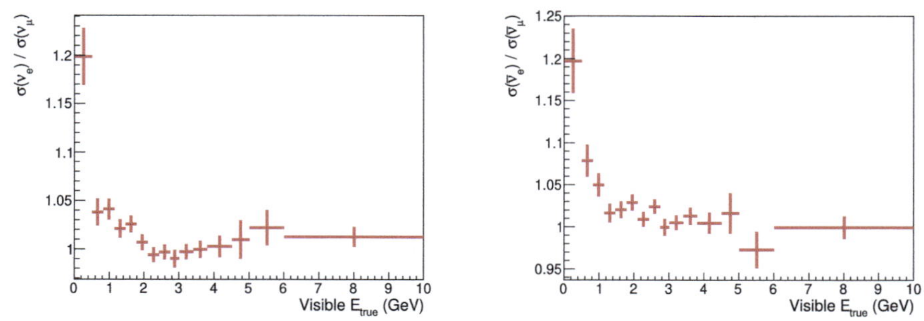

(a) $\sigma(\nu_e)/\sigma(\nu_\mu)$ (b) $\sigma(\bar{\nu}_e)/\sigma(\bar{\nu}_\mu)$

Fig. 4.11 Cross section ratios for the electron neutrino appearance analysis in FHC and RHC mode as a function of $E^{vis.}_{true}$. The error bars originate from the statistical uncertainty on the MC predictions used to calculate the ratios

4.3.4 Introduce Far Detector Effects

The final step in this section is to account for the resolution and selection efficiency in the FD for each signal channel: ν_μ, $\bar{\nu}_\mu$, ν_e and $\bar{\nu}_e$. Like the efficiency calculation in the ND, the selection efficiency in the FD is calculated for the four signal channels of interest as a function of $E^{true}_{vis.}$. Whilst the ND cuts were mostly acceptance cuts derived from the detector geometry, almost all events that occur in the fiducial volume of each FD module are fully contained. Therefore, the selection efficiency in the FD is driven entirely by the performance of the CVN. As can be seen in the selection efficiencies for ν_μ, $\bar{\nu}_\mu$, ν_e and $\bar{\nu}_e$ shown in Fig. 4.12, the CVN achieves efficiencies of the order 90% around the main oscillation region of 1–5 GeV.

To correct the ND data for the FD selection efficiency and resolution another set of smearing matrices are created that map the jth true visible energy bin E^{true}_j to the ith reconstructed neutrino energy bin E^{reco}_i in the FD. The reconstructed energy axis is the $E^{FD}_{rec.}$ variable calculated from Eq. 4.1. For the disappearance analysis the matrix is populated with ν_μ events for FHC mode and $\bar{\nu}_\mu$ events in RHC mode. The appearance measurement populates the smearing matrices with ν_e events for FHC mode and $\bar{\nu}_e$ events in RHC mode. The matrix $\mathbf{M^{FD}}$ is normalised to the FD selection

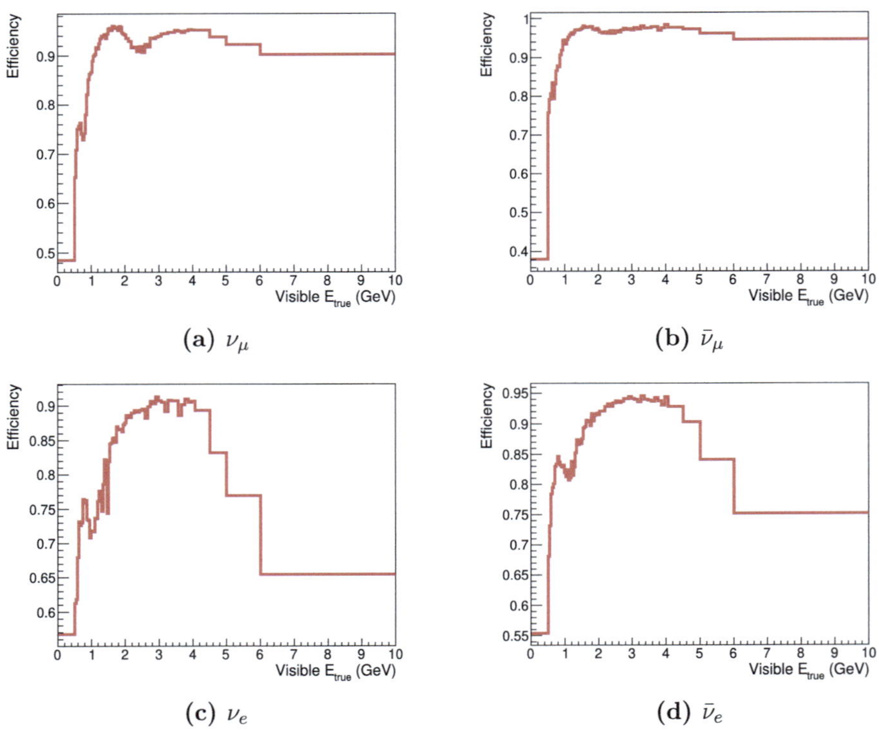

Fig. 4.12 FD ν_μ, $\bar{\nu}_\mu$, ν_e and $\bar{\nu}_e$ selection efficiency calculated as a function of $E^{vis.}_{true}$

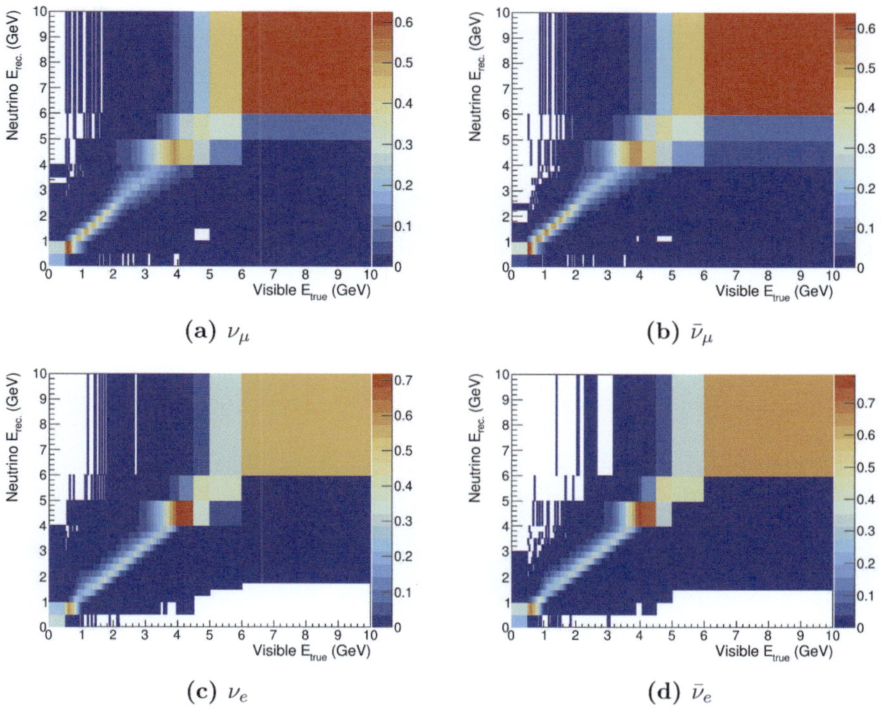

Fig. 4.13 FD ν_μ, $\bar{\nu}_\mu$, ν_e and $\bar{\nu}_e$ smearing matrices normalised to the selection efficiencies shown in Fig. 4.12

efficiency as a function of true visible energy in same way as Eq. 4.6. Examples of the FD ν_μ, $\bar{\nu}_\mu$, ν_e and $\bar{\nu}_e$ smearing matrices are shown in Fig. 4.13. Each slice of unfolded ND data is smeared by the normalised FD matrix such that

$$E_i^{rec.} = \sum_j^{n_{true}} M_{ij}^{FD} E_j^{true}. \tag{4.14}$$

Smearing each slice of unfolded ND data according to Eq. 4.14 provides samples of ND data that are a function of $E_{rec.}^{FD}$ and off-axis position that have been corrected for ND backgrounds, selection efficiency and resolution differences between the ND and FD. The corrections are applied for the signal channel of interest. For example, if a data-driven prediction of the $\nu_\mu \rightarrow \nu_e$ appearance spectrum at the FD is desired, then the FD effects introduced are for ν_e selection efficiency and resolution in the FD.

The final covariance matrix for each off-axis position $\mathbf{V}^{\mathbf{smr.}}$ is propagated through the FD smearing procedure. The covariance matrix for the unfolded solution $\mathbf{V}^{\mathbf{unf.}}$

is calculated for each off-axis position in Eq. 4.13. The statistical uncertainty on the smeared ND data at a particular off-axis position is then

$$\mathbf{V}^{\text{smr.}} = \mathbf{M}^{\text{FD}}\mathbf{V}^{\text{unf.}}\left(\mathbf{M}^{\text{FD}}\right)^{\text{T}}. \tag{4.15}$$

Figure 4.14 shows the corrected ND data that can be used in a linear combination PRISM measurement of $\nu_\mu \rightarrow \nu_\mu$ (Fig. 4.14a), $\bar{\nu}_\mu \rightarrow \bar{\nu}_\mu$ (Fig. 4.14b), $\nu_\mu \rightarrow \nu_e$ (Fig. 4.14c) and $\bar{\nu}_\mu \rightarrow \bar{\nu}_e$ (Fig. 4.14d) oscillations. The unfolding and smearing procedure detailed in this Section provides a sample of ND event rates at many off-axis positions that have the detector characteristics of the FD. In addition, this procedure propagates the statistical uncertainty on the ND data, which can be correctly accounted for in the linear combination technique described in the following sections.

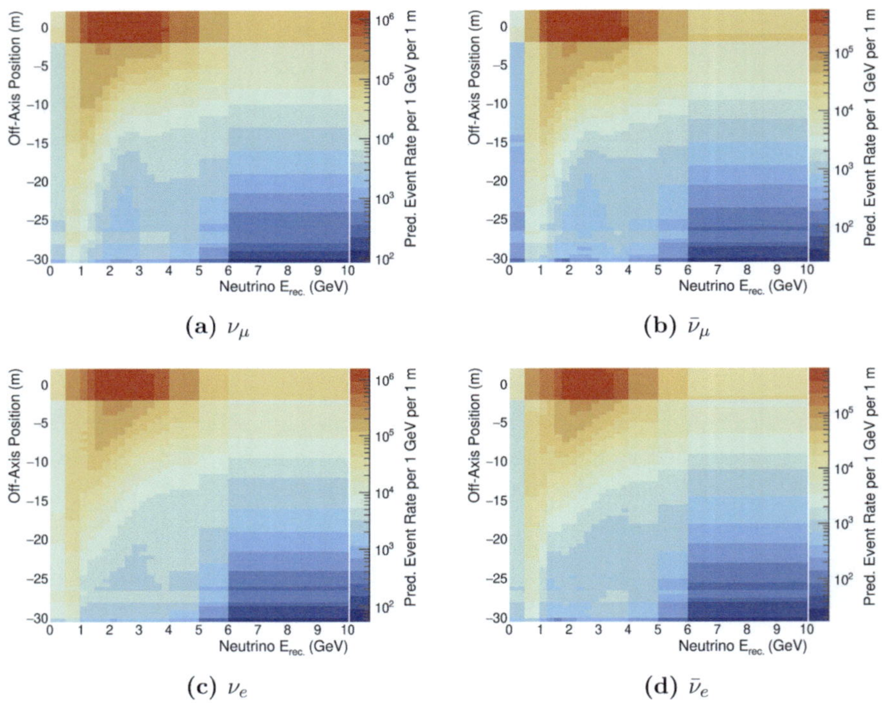

Fig. 4.14 ND data that has corrections applied to account for differences between the ND and FD

4.4 Linear Combination

The ND data corrected for backgrounds and detector differences at each off-axis position for the two horn polarities and currents can be linearly combined to produce a data-driven prediction of the DUNE FD event rate. Each off-axis bin i is summed with a corresponding coefficient c_i to produce a linearly combined FD prediction:

$$F_j^{LC} = \sum_i^{n_{pos}} N_{ij}^{data} c_i, \tag{4.16}$$

where N_{ij}^{data} are matrix elements corresponding to the ND event rate in ith off-axis position bin and jth reconstructed energy bin. The task then is to calculate what coefficients results in a linear combination that closely matches the FD event rate. These coefficients are calculated from the flux model, entirely *independent* of the neutrino interaction model. True ν_μ and $\bar{\nu}_\mu$ CC event rates as a function of the true neutrino energy from the ND and FD MC data are used for this calculation, which are simply $flux \times cross\ section$. The cross section for a neutrino interacting with a nucleon in an argon nucleus as a function of neutrino energy is identical in the ND and FD and so cancels when mapping the ND event rates to the FD oscillated event rate. Therefore, calculating the linear combination coefficients using the true CC event rates is still independent of the neutrino interaction model. For convenience, the true CC event rates are referred to in this section as the "flux", since the cross section is simply a scaling factor that cancels in the procedure described below.

The flux simulation provides a matrix of the ND fluxes $\mathbf{N}$ and a vector of the oscillated FD flux F. Section 4.1.4 discussed how an additional on-axis ND data sample measured at a lower horn current is also added to the linear combination. The ND flux matrix therefore includes both the 293 kA and 280 kA flux samples. The oscillated FD flux vector depends on oscillation parameters and the oscillation transition of interest. For example, when building a PRISM prediction of the $\nu_\mu \to \nu_e$ appearance event rate F is the ν_e appearance flux for the chosen oscillation parameters. True ν_μ CC event rates are also used to produce the appearance flux shape for F in order for the ν_μ CC cross section to cancel when mapping the ND ν_μ CC event rates to F.

A linear algebra equation of the form $\mathbf{N}c = F$ is solved for c, which are the linear combination coefficients to be used in Eq. 4.16. This is an ill-posed linear algebra problem. If the equation were to be naively solved by matrix inversion fluctuations in the FD flux prediction would lead to large variations in a highly degenerate solution. Instead, the method of Tikhonov regularisation is again used to find a stable approximate solution, where variations between adjacent elements of c are dampened. The solution for c is calculated by minimising

$$\|\mathbf{N}c - F\|^2 + \|\Gamma c\|^2, \tag{4.17}$$

where Γ is the regularisation matrix, which forces the adjacent elements of c to be similar. The strength of the regularisation is determined by a parameter, τ_{LC}, and the regularisation matrix has the form

$$\Gamma = \tau_{LC} \begin{pmatrix} 1 & -1 & 0 & 0 & \cdots & 0 & 0 \\ 0 & 1 & -1 & 0 & \cdots & 0 & 0 \\ 0 & 0 & 1 & -1 & \cdots & 0 & 0 \\ 0 & 0 & 0 & 1 & \cdots & 0 & 0 \\ \vdots & \vdots & \vdots & \vdots & \ddots & \vdots & \vdots \\ 0 & 0 & 0 & 0 & \cdots & 1 & -1 \\ 0 & 0 & 0 & 0 & \cdots & 0 & 0 \end{pmatrix}. \tag{4.18}$$

Similarly to Eqs. 4.11 and 4.12, the linear combination coefficients are calculated from the expression

$$c = \left((\mathbf{N})^{\mathrm{T}} \mathbf{PN} + \Gamma^{\mathrm{T}} \Gamma \right)^{-1} (\mathbf{N})^{\mathrm{T}} \mathbf{P} F, \tag{4.19}$$

where $\mathbf{P}$ is a diagonal matrix used to down-weight regions of neutrino energy where it is difficult to achieve a good mapping between the ND fluxes and the FD flux. This prevents the lowest and highest neutrino energies (below 0.5 GeV and above 5 GeV) from affecting the coefficient calculation and prioritises achieving a good mapping between $\mathbf{N}$ and F between 0.5 and 5 GeV, which is the main oscillation region for DUNE.

Examples of solutions for c in the four signal channels can be seen in Fig. 4.15. Notice how the coefficients all vary smoothly. This is a consequence of the regularisation condition. How smooth the coefficients are is determined by the regularisation parameter τ_{LC}. A sensible choice for this parameter is important for PRISM since the smoothness of the linear combination coefficients controls the size of the statistical uncertainties in the analysis. Hence, an optimisation of this parameter is performed using the so-called "L-curve" method and more details on this can be found in Appendix A. The ν_μ ($\bar{\nu}_\mu$) weights match the ND ν_μ ($\bar{\nu}_\mu$) off-axis fluxes to the FD ν_μ ($\bar{\nu}_\mu$) flux for the chosen set of oscillation parameters. Likewise, the ν_e ($\bar{\nu}_e$) weights match the ND ν_μ ($\bar{\nu}_\mu$) off-axis fluxes to the FD ν_e ($\bar{\nu}_e$) flux. Linear combination coefficients can therefore be calculated for any possible set of oscillation parameters by changing the oscillation hypothesis assumed for the FD flux.

4.5 Far Detector Monte Carlo Corrections

The backgrounds at the ND are subtracted away. Therefore, the predicted backgrounds at the FD must be added to the linearly combined ND data. The FD backgrounds that are added to the final prediction can be seen in Fig. 4.3. This is a correction that comes entirely from the FD MC. However, it is possible for some of these backgrounds to have data-driven predictions using the same linear combi-

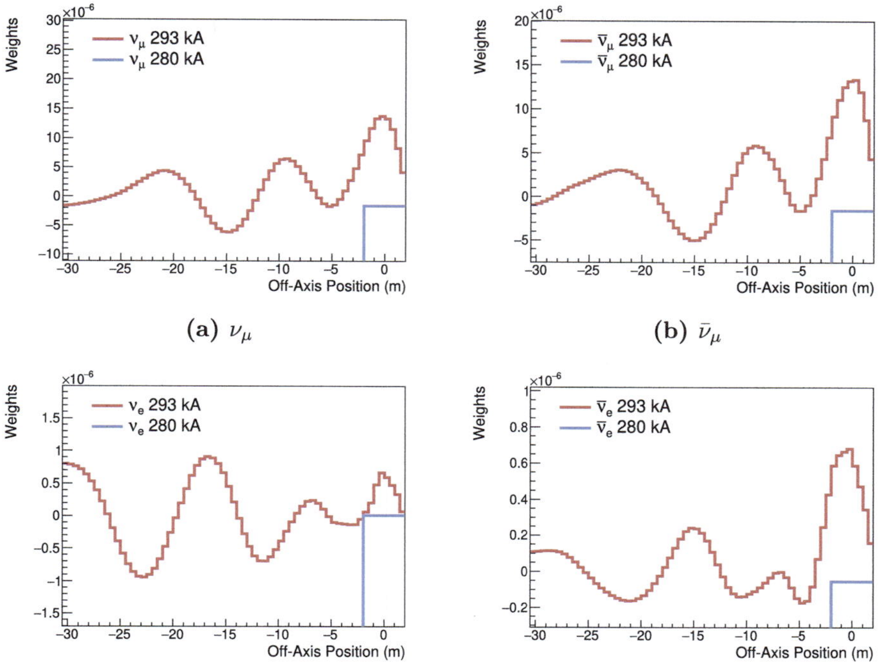

(a) ν_μ (b) $\bar{\nu}_\mu$

(c) ν_e (d) $\bar{\nu}_e$

Fig. 4.15 Linear combination coefficients calculated by matching ND off-axis fluxes to the FD flux. Shown are the coefficients for the 293 kA and 280 kA ND flux samples

nation method described. At the end of this chapter Sect. 4.7.2 provides a detailed discussion on how data-driven methods for prediction the FD backgrounds will be implemented in a future PRISM oscillation analysis.

An additional correction from the FD MC is required to correct for imperfections in linear combination coefficient calculation. Muon neutrino fluxes at the ND have a maximum peak at approximately 2.5 GeV in the on-axis position and the high-energy tail in the flux becomes very small above 5 GeV. Above this neutrino energy it is almost impossible to match the ND off-axis fluxes to the FD oscillated flux. In addition, although the regularisation provides a set of linear combination coefficients with desirable statistical properties, the consequence of this is the introduction of a small amount of bias to the solution. If the quantity $\mathbf{N}c$ is calculated using the set of coefficients derived using Tikhonov regularisation (Eq. 4.19), it can be seen in Fig. 4.16 that $\mathbf{N}c$ does not perfectly match the FD flux. The fractional difference between $\mathbf{N}c$ and the FD flux is called the residual, which is used to calculate the FD flux-matching correction.

The FD flux-matching correction is calculated from a FD MC prediction that has two axes: true neutrino energy versus $E_{FD}^{rec.}$. The true neutrino energy axis is weighted by the residual of linear combination coefficient calculation seen in

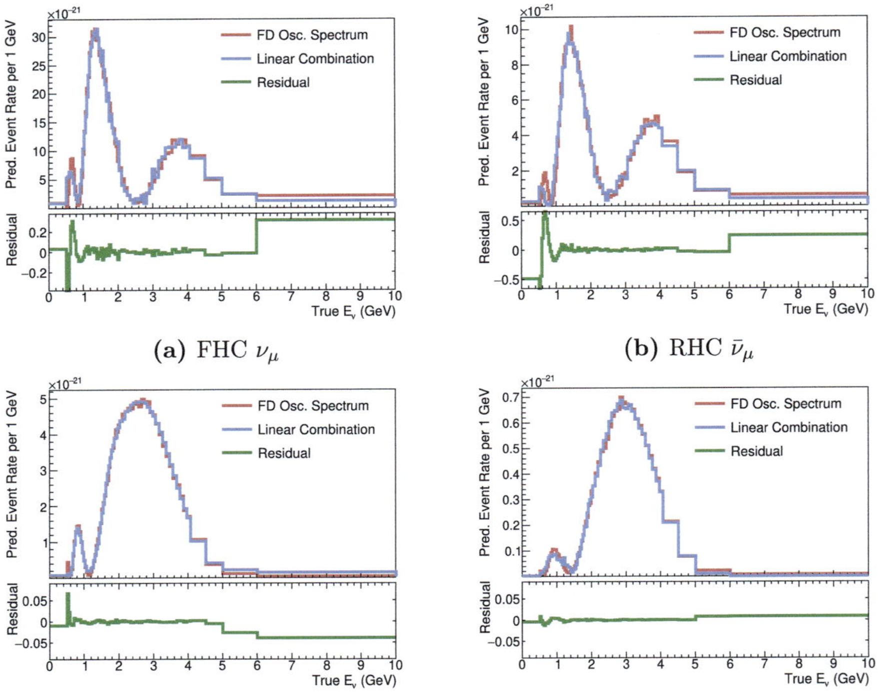

(a) FHC ν_μ (b) RHC $\bar{\nu}_\mu$ (c) FHC ν_e (d) RHC $\bar{\nu}_e$

Fig. 4.16 FD flux prediction and the linearly combined ND fluxes. The residual is the fractional difference between the FD flux and the linearly combined ND fluxes

Fig. 4.16. This gives a FD flux-matching correction as a function of reconstructed neutrino energy that can be calculated for any possible set of linear combination coefficients. Figure 4.17 shows the stacked combination of the FD backgrounds and the FD flux-matching correction to be added to the linear combination of ND data in order to build the total PRISM prediction for the four oscillation channels.

4.6 PRISM Predictions

The final PRISM prediction is the sum of the linear combination of ND data, the FD backgrounds B_j and the FD flux-matching correction R_j:

$$F_j^{PRISM} = \sum_i^{n_{pos}} N_{ij}^{data} c_i + B_j + R_j, \tag{4.20}$$

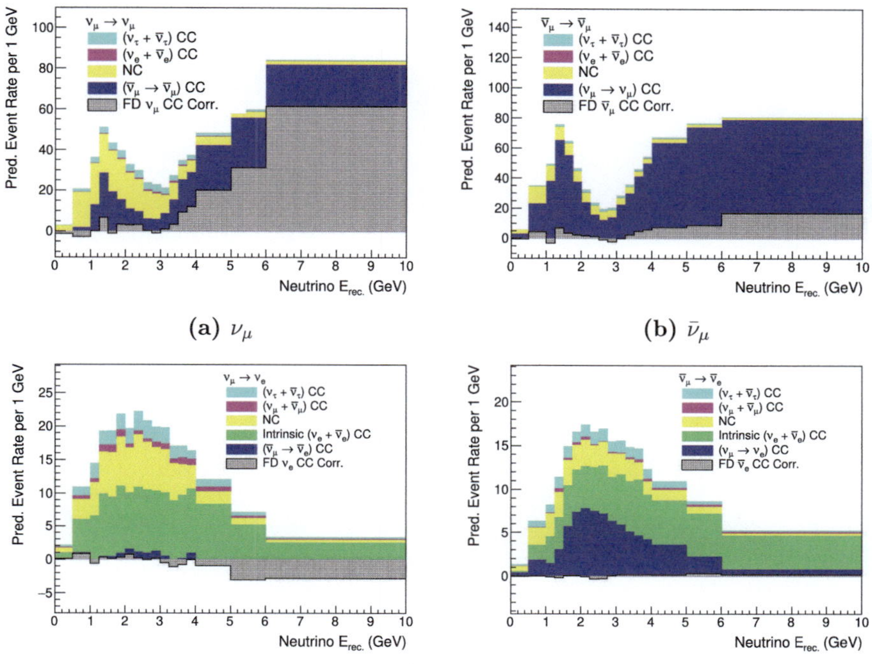

(a) ν_μ

(b) $\bar{\nu}_\mu$

(c) ν_e

(d) $\bar{\nu}_e$

Fig. 4.17 Corrections from the FD MC that include backgrounds and the flux miss-match correction. The FD exposure is 50 kt-MW-Yrs

where i and j are bins of off-axis position and $E_{FD}^{rec.}$ respectively. The linear combination coefficients are calculated from flux model, following the method explained in Sect. 4.4. The coefficients, like those shown in Fig. 4.15, are weighted to account for the particular ND run-plan. For the results presented in this thesis, this means weighting down the on-axis coefficients according to the run-plan shown in Sect. 4.2. The weighted linear combination coefficients are applied in a linear sum to each slice of corrected off-axis ND data shown in Figure 4.14. The final PRISM predictions in FHC and RHC mode for the four DUNE oscillation signal channels, assuming the NuFIT 4.0 oscillation parameters [3], are shown in Fig. 4.18. The stacked histograms in Fig. 4.18 are the total PRISM prediction, where the red component is the linear combination of ND off-axis data.

As discussed in Sect. 4.3.2, the procedure to correct the ND data for efficiency and resolution differences between the ND and FD can introduce correlations between the bins of off-axis data. Consequently, each off-axis position of ND data has a covariance matrix associated with it. Since each off-axis slice of ND data is linearly combined, the covariance matrices at each off-axis position are also combined with the corresponding coefficient-squared. This calculation of the PRISM covariance matrix can be expressed

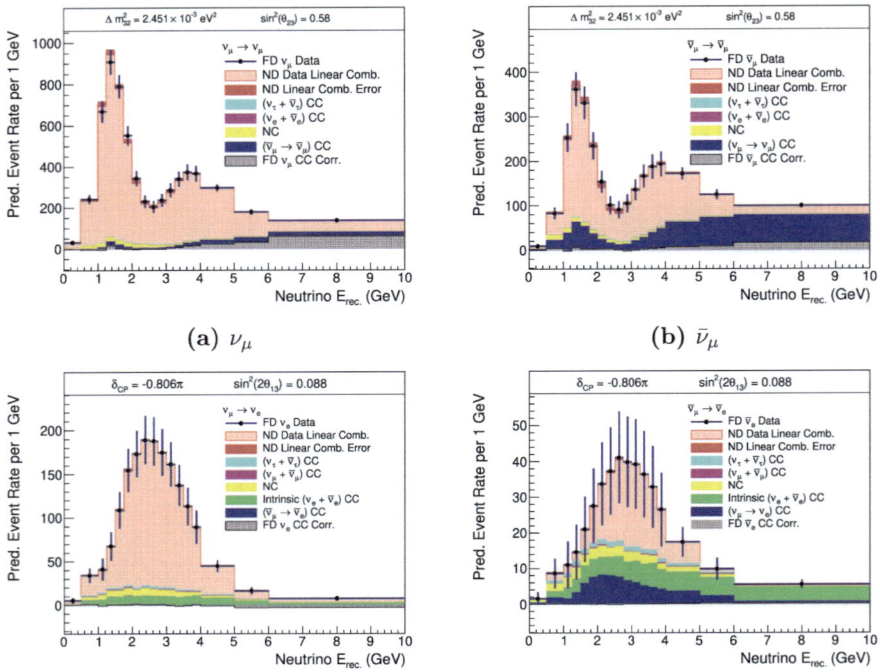

(a) ν_μ (b) $\bar{\nu}_\mu$

(c) ν_e (d) $\bar{\nu}_e$

Fig. 4.18 PRISM predictions corresponding to set of linear combination coefficients shown in Fig. 4.15 for the four DUNE signal channels mode. The total stacked histogram is the PRISM prediction. The FD exposure is 50 kt-MW-Yrs

$$\mathbf{V}^{\mathbf{PRISM}} = \sum_{i}^{n_{pos}} c_i^2 \mathbf{V}^{\mathbf{smr.}}{}_i, \tag{4.21}$$

where $\mathbf{V}^{\mathbf{smr.}}{}_i$ is the covariance matrix for the unfolded and smeared ND data at off-axis position i calculated from Eq. 4.15. This gives a final covariance matrix for the PRISM prediction and Fig. 4.19 shows the covariances associated with the PRISM predictions seen Fig. 4.18. The square-root of the diagonal elements of the covariance matrices shown in Fig. 4.19 correspond exactly to the linear combination error bars plotted on the PRISM predictions in Fig. 4.18.

Equation 4.21 also highlights the importance of making a good choice for the linear combination regularisation parameter τ_{LC}. If the regularisation is too weak when mapping the ND fluxes to the FD flux, the calculated linear combination coefficients c will oscillate between extreme positive and negative values. The magnitude of the final covariances on the PRISM prediction depend on c_i^2, so a one order of magnitude increase in the average size of the coefficient will lead to a two orders of magnitude increase in the covariances.

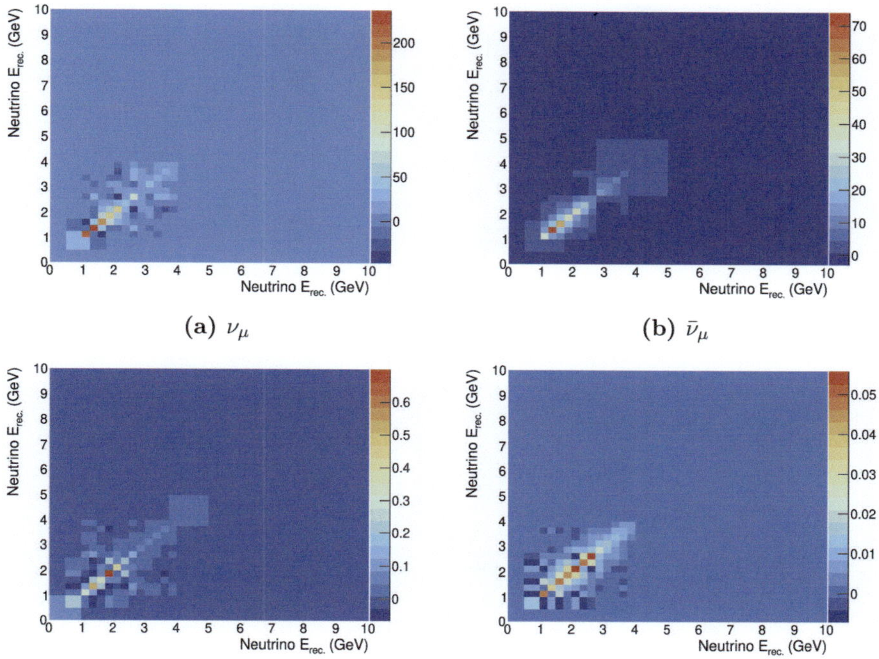

(a) ν_μ (b) $\bar{\nu}_\mu$

(c) ν_e (d) $\bar{\nu}_e$

Fig. 4.19 Covariance matrices for ν_μ, $\bar{\nu}_\mu$, ν_e and $\bar{\nu}_e$ predictions. The covariance matrices are associated with the PRISM predictions shown in Fig. 4.18. The diagonal elements are the variance of each bin of the PRISM prediction

4.6.1 Predictions for any Oscillation Parameter Hypothesis

So far, the NuFIT 4.0 oscillation parameters have been assumed. However, PRISM predictions of the FD event rate can be made for any set of oscillation parameters. The oscillation parameters of the PRISM prediction can be changed by simply altering the oscillation parameters of the FD flux prediction when calculating the linear combination coefficients. Predictions for the background contributions to the PRISM prediction are taken from the MC and so can also be calculated for any oscillation hypothesis.

The parameters Δm^2_{32} and θ_{23} dominate the oscillation probability for muon neutrino disappearance. In Fig. 4.20 it is shown how ν_μ and $\bar{\nu}_\mu$ disappearance predictions can be made for any combination of Δm^2_{32} and θ_{23} whilst still matching the FD data well across the whole energy range. Measurements of electron neutrino appearance at the FD is sensitive to the parameters θ_{13} and δ_{CP}. Figure 4.21 demonstrates how ν_e and $\bar{\nu}_e$ appearance PRISM predictions can match the FD data well for any set of θ_{13} and δ_{CP} values.

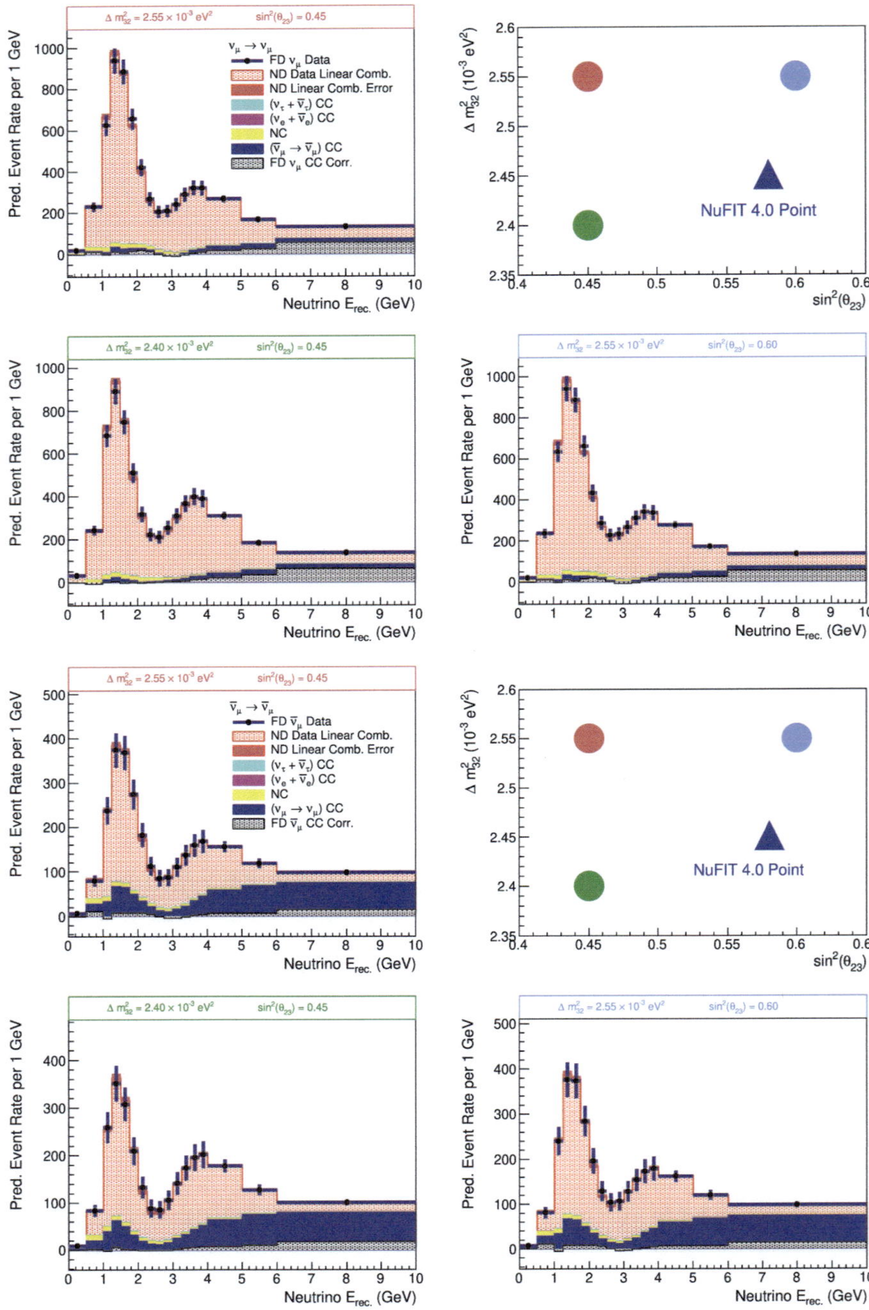

Fig. 4.20 PRISM disappearance predictions for three different oscillation hypotheses in FHC (top) and RHC (bottom) mode. Muon neutrino disappearance is dominated by Δm^2_{32} and $\sin^2 \theta_{23}$ and the different values of these parameters for the three predictions are shown in the oscillation parameter space at the top right of each plot

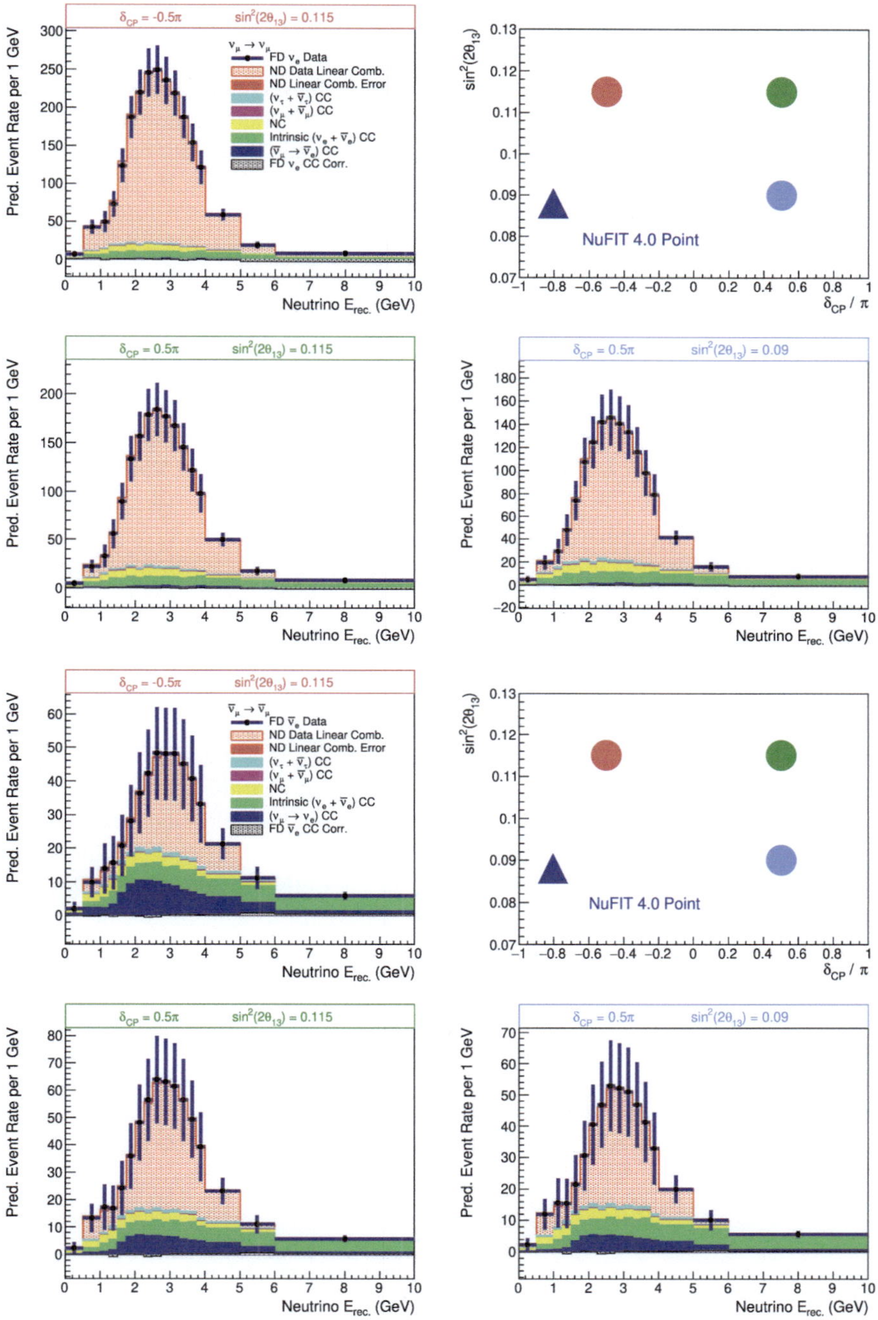

Fig. 4.21 PRISM appearance predictions for three different oscillation hypotheses in FHC (top) and RHC (bottom) mode. Measurements of electron neutrino appearance are sensitive to δ_{CP} and $\sin^2 2\theta_{13}$ and the different values of these parameters for the three predictions are shown in the oscillation parameter space at the top right of each plot

Figures 4.20 and 4.21 provide a preview of how the PRISM linear combination method can be used to measure the oscillation parameters. An oscillation parameter space can be defined and a PRISM prediction made for any point in that parameter space. At each point in the oscillation parameter space the goodness-of-fit of the PRISM prediction to the data can be calculated, allowing sensitivity contours to be drawn. Oscillation sensitivity contours can then be made for Δm^2_{32}, θ_{23}, θ_{13} and δ_{CP}, which are all the oscillation parameters that will be measured by DUNE.

This chapter has focused on the methodology of producing data-driven oscillated FD event rate predictions in a PRISM analysis. As will be shown in Chap. 6, these predictions provide DUNE with sensitivity to all the oscillation parameters in the three-flavour oscillation model accessible by long baseline neutrino oscillation experiments. The methodology presented in this chapter for producing a PRISM prediction should not be regarded as final. There are many improvements that can be made to the PRISM methodology and the final section of this chapter highlights some of the features that will be added to a future PRISM analysis. These improvements aim to mitigate the impact of systematic uncertainties on the analysis through the various model-dependent components of the PRISM prediction, which will be discussed in detail in Chap. 5.

4.7 Future Methodology Improvements

The methodology for producing a PRISM prediction of the DUNE FD event rate has been explained in detail in this chapter. The purpose of the PRISM method is to reduce the neutrino interaction model dependence of the oscillation measurement by building a prediction of the FD event rate from data, not from the MC simulation. Whilst a method has been implemented that successfully predicts the FD event by linearly combining off-axis ND data, this is achieved through the use of the MC simulation at several steps in the methodology. Components of the prediction that are derived from the MC are dependent on the neutrino interaction model and will be affected by systematic uncertainties. The MC is used to correct for differences between near and far detector effects (Sect. 4.3) and to account for backgrounds and a flux-matching correction (Section 4.5). As will be demonstrated in later chapters, the PRISM analysis can be improved significantly if some of these MC components can be replaced by data-driven methods. This section will summarise the plans for a future PRISM analysis that will include such data-driven methods.

4.7.1 Geometric Efficiency Correction

Differences between the event selection efficiencies at the near and far detectors are currently corrected for using the MC. The performance of the selection cuts in the ND and FD are very different due to the different acceptances of the detectors. Each

FD module is a single large body of liquid argon in which almost all neutrino events are well-contained. The selection efficiency in the FD is therefore driven entirely by the performance of the CVN, which, as seen in Fig. 4.12, provides efficiencies of the order 90% across most energies and in each of the four signal channels. Since the total and selected event rates in the FD are similar, it is expected that shifts in the MC predictions due to systematic uncertainties will cancel well in the calculation of the FD selection efficiency.

The situation is quite different in the ND selection, where the overall selection efficiency is significantly lower than in the FD due to the acceptance cuts that require the muon to be either contained in ND-LAr or to propagate to TMS/ND-GAr, as well as the hadronic containment cut. Figure 4.10 demonstrates that the ND selection efficiency can be as low as 20% in some regions of phase space. Correcting for such low efficiencies using only the MC risks introducing significant model dependence into the PRISM analysis. Hence, a data-driven method for correcting for the ND acceptance cuts—known as the "geometric efficiency correction"—is being developed for PRISM. The "geometric efficiency" refers to the efficiency of the acceptance cuts.

The ND acceptance cuts are separated into the muon and hadronic shower cuts, both of which must be passed for a neutrino event to be selected in the ND. A neutrino event should have rotational symmetry around the axis of the incoming neutrino. In addition, the distribution of neutrino events across the y–z plane should be flat, since the neutrino flux is not expected vary substantially in the y or z direction on the scale of the ND-LAr active volume. Assuming these two conditions are true, for each selected ND event many random rotations about the incoming neutrino axis and translations in the y–z plane at the off-axis position of the event are performed. For each random rotation and translation, the hadronic shower either passes or fails the hadronic veto cut and the muon is assigned a probability quantifying how likely it is that it would be either contained in ND-LAr or propagate to TMS/ND-GAr. The probability of the muon passing the acceptance cuts is calculated by a neural network that is trained to determine the likelihood of muon being selected given its initial kinematics. The fraction of random throws that pass both the hadronic containment and muon cuts provides the geometric efficiency for the selected ND event. In other words, this procedure quantifies how probable it is, given the ND acceptance, that a particular ND event is selected on an event-by-event basis. The ND efficiency is then corrected for by weighting each event by one over its calculated probability of being selected. Figure 4.22 provides a sketch that illustrates how the geometric efficiency is calculated for an event.

A complicating factor in the method described above is the presence of events in the FD that are never selected in the ND. For example, events that have a very large hadron shower could always fail the hadronic veto cut regardless of how the geometric transformations are applied to the event. A data-driven ND efficiency correction cannot correct for events that are never selected in the ND. The proposed strategy to handle such events is as follows:

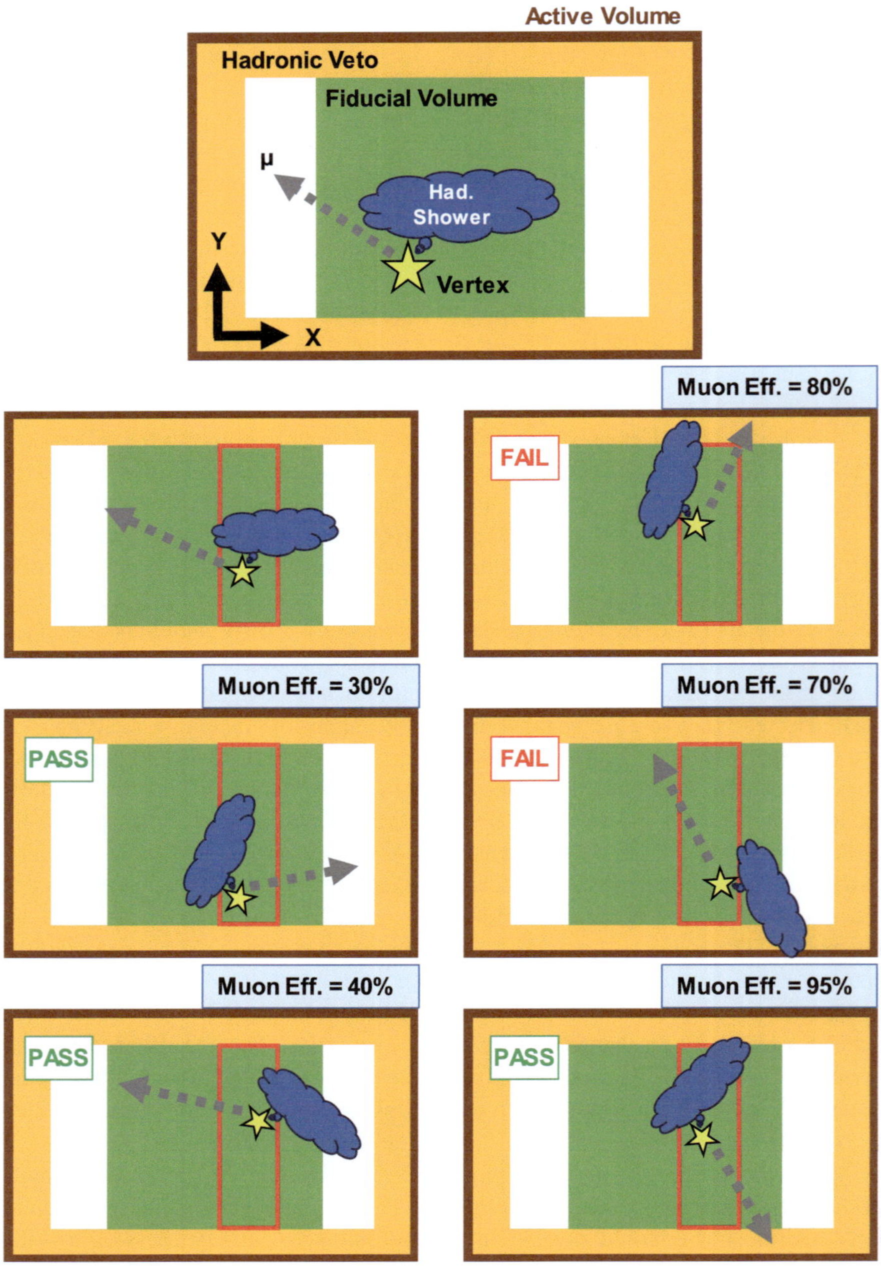

Fig. 4.22 Sketch illustrating the geometric efficiency calculation. The neutrino event is rotated about the neutrino axis and translated in the y–z plane a number of times. The red box represents the fixed x position of the y–z plane in which the event is translated. Whether the hadronic veto cut is passed or failed is indicated. The muon selection probability is also shown. For the five random throws shown the geometric efficiency is therefore: $(0 \times 0.8 + 1 \times 0.3 + 0 \times 0.7 + 1 \times 0.4 + 1 \times 0.95)/5 = 0.33$

1. Identify FD data events with large hadron showers and transfer their energy deposits to ND-LAr.
2. Calculate the ND geometric efficiency for the FD events.
3. Remove the FD events from the FD signal sample that have a geometric efficiency below a given threshold.

At the time of writing, the geometric efficiency correction is under development by the DUNE-PRISM working group within the DUNE Collaboration. When implemented, a data-driven and model-independent method will be in-place to correct for the ND geometric efficiency. The MC will still be relied upon to correct for any non-acceptance cuts, such as the CVN in the FD, or a more sophisticated signal selection in the ND. However, it is the acceptance cuts that most-severely degrade the total selection efficiency at the ND. Consequently, it is expected that the implementation of this data-driven method will result in a substantial reduction in the model dependence of the PRISM analysis.

4.7.2 Linear Combination Predictions of the Far Detector Backgrounds

The linear combination method detailed in this chapter has focused on linearly combining off-axis ND data to predict the signal events at the FD. Only predicting the signal at the FD means the FD backgrounds are derived from the MC, which introduces model dependence. Whilst the linear combination typically forms the largest component, the MC background predictions do constitute a significant contribution to the overall PRISM prediction, particularly for the RHC signal channels. It is possible to use the same linear combination method to build data-driven predictions of certain FD background components, thereby reducing the reliance on the MC simulation.

Figures 4.20 and 4.21 show that the FD wrong-sign background is a significant contribution to the RHC PRISM predictions. It is possible to reduce the neutrino interaction model dependence of this component by producing a data-driven prediction of the wrong-sign background. This is done by linearly combining ND ν_μ events measured in FHC mode to predict the FD RHC $\nu_\mu \rightarrow \nu_\mu$ wrong-sign background. The same methodology discussed in Sect. 4.3 for correcting for efficiency and resolution differences between the ND and FD will need to be employed when matching ND FHC ν_μ events to FD RHC ν_μ. Producing a data-driven prediction of the wrong-sign background in RHC mode could significantly reduce the neutrino interaction model dependence in the $\bar{\nu}_\mu \rightarrow \bar{\nu}_\mu$ prediction.

The FD ν_e and $\bar{\nu}_e$ selection is unable to distinguish between electron neutrinos that appeared in the beam through oscillations ($\nu_\mu \rightarrow \nu_e$) and those that are intrinsic to the beam ($\nu_e \rightarrow \nu_e$). Figure 4.21 shows that the intrinsic $\nu_e + \bar{\nu}_e$ background is a significant contribution to the PRISM appearance predictions. Currently, the methodology calculates the appearance linear combination coefficients by matching

ND ν_μ ($\bar{\nu}_\mu$) fluxes to the FD appearance $\nu_\mu \rightarrow \nu_e$ ($\bar{\nu}_\mu \rightarrow \bar{\nu}_e$) flux. However, there is no reason why the ND ν_μ ($\bar{\nu}_\mu$) fluxes cannot be matched to the combined appearance and intrinsic ν_e ($\bar{\nu}_e$) flux. Doing so provides a set of linear combination coefficients that incorporates the right-sign intrinsic ν_e ($\bar{\nu}_e$) background into the PRISM $\nu_\mu \rightarrow \nu_e$ ($\bar{\nu}_\mu \rightarrow \bar{\nu}_e$) linear combination of ND data. Visually, this means a large component of the green stacked histogram in the predictions shown in Figure 4.21 is incorporated into the red linear combination component. Only the right-sign intrinsic ν_e ($\bar{\nu}_e$) is included in the linear combination due to the need to apply the cross section ratio correction described in Sect. 4.3.3. The ratio $\sigma(\nu_e)/\sigma(\nu_\mu)$ is not necessarily the same as $\sigma(\bar{\nu}_e)/\sigma(\nu_\mu)$.

4.7.3 PRISM $\sigma(\nu_e)/\sigma(\nu_\mu)$ Measurement

The PRISM electron neutrino appearance prediction linearly combines ND ν_μ ($\bar{\nu}_\mu$) data to predict the FD ν_e ($\bar{\nu}_e$) event rate. As explained in Sect. 4.3.3, this requires a correction for the cross section ratio $\sigma(\nu_e)/\sigma(\nu_\mu)$ that is currently derived from the MC predictions of ν_μ ($\bar{\nu}_\mu$) and ν_e ($\bar{\nu}_e$) event rates. It is possible that PRISM could perform a measurement of this ratio using off-axis data. In general, the PRISM method involves linearly combining ND fluxes to match some target flux. This chapter has focused on the case where the target flux is the FD oscillated signal. However, in principle any flux shape can be created from the linear combination.

One possibility is to linearly combine ND ν_μ ($\bar{\nu}_\mu$) data to match a sample of ND ν_e ($\bar{\nu}_e$) event rates. In this case the fluxes of the linearly combined ν_μ ($\bar{\nu}_\mu$) data and the ν_e ($\bar{\nu}_e$) sample will be the same, meaning any differences between the two event rates must be due to the cross section ratio. This method is complicated by the need to correct for selection efficiency and resolution differences between ν_μ and ν_e in the ND. However, provided such corrections can be included, perhaps using a method similar to that described in Sect. 4.3, this technique would provide a direct measurement of $\sigma(\nu_e)/\sigma(\nu_\mu)$ rather than simply relying on the MC.

4.7.4 Alternative Analysis Variables

The default PRISM method uses a one-dimensional reconstructed neutrino energy variable to perform the oscillation analysis. However, it may be advantages to implement a different analysis variable that predicts the FD event rate in more than one dimension. For example, T2K has used both an $E^{rec.}$ versus $\theta_{lep.}$ and a $p_{lep.}$ versus $\theta_{lep.}$ set of axes in its oscillation analysis, where $E^{rec.}$ is the reconstructed neutrino energy, $\theta_{lep.}$ is the angle between the lepton and the neutrino beam and $p_{lep.}$ is the lepton momentum. This choice assists T2K in separating neutrinos and antineutrinos in its oscillation analysis [11].

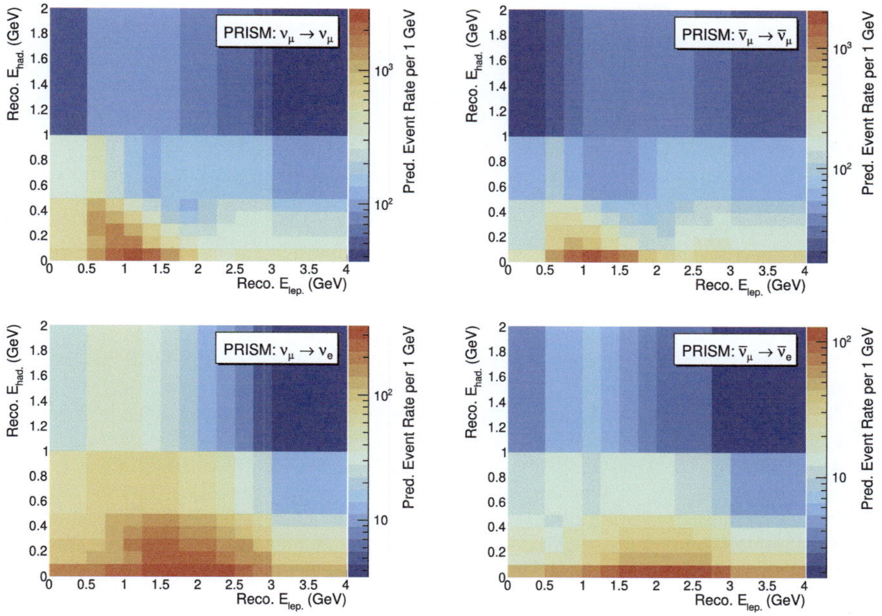

Fig. 4.23 Total PRISM predictions of the FD event rate using the two-dimensional analysis variable $E_{had.}$ versus $E_{lep.}$

Similarly, any choice of analysis variable can be made for the PRISM method. Figure 4.23 demonstrates how a PRISM analysis can be done with the reconstructed lepton and hadronic energy plotted against one another on separate axes. The same methodology outlined in this chapter is followed for producing a PRISM prediction with any choice of analysis variable. In principle, this choice of analysis variable would make the smearing matrices used to correct for detector differences between the ND and FD (Sect. 4.3) less dependent on the neutrino interaction model. This is because events with a large lepton energy and small hadronic energy and vice versa, which could have a very different cross sections, no longer occupy the same reconstructed neutrino energy bin. However, this variable choice is found to be a weak measure of the oscillation probability distribution, degrading the sensitivity of the PRISM oscillation measurement. A two-dimensional analysis variable is also much more computationally expensive, essentially squaring the number of analysis bins that need to be propagated through the PRISM methodology described in this chapter. Further study is needed to determine whether an alternative two-dimensional analysis variable could benefit a PRISM oscillation analysis.

References

1. DUNE Collaboration, B. Abi, et al., Deep underground neutrino experiment (DUNE), far detector technical design report, in *DUNE Physics*, vol. II (Feb 2020). arXiv: 2002.03005 [hep-ex]
2. DUNE Collaboration, B. Abi, et al., Neutrino interaction classification with a convolutional neural network in the DUNE far detector. Phys. Rev. D, **102**(9), 092 003 (2020). arXiv: 2006.15052 [physics.ins-det]
3. I. Esteban et al., Global analysis of three-flavour neutrino oscillations: synergies and tensions in the determination of θ_{23}, δ_{CP}, and the mass ordering. JHEP **01**(106) (2019). arXiv: 1811.05487 [hep-ph]
4. DUNE Collaboration, A.A. Abud, et al., Deep underground neutrino experiment (DUNE) near detector conceptual design report. Instruments **5**(4), 31 (2021). arXiv: 2103.13910 [physics.ins-det]
5. D.L. Phillips, A technique for the numerical solution of certain integral equations of the first kind. J. Assoc. Comput. Mach. **9**(1), 84–97 (1962)
6. A.N. Tikhonov, V.Y. Arsenin, *Solutions of Ill-Posed Problems*, vol. xiii258. (John Wiley & Sons, Washington, D.C., V. H. Winston & Sons, New York, 1977). (Translated from the Russian, Preface by translation editor Fritz John, Scripta Series in Mathematics)
7. S. Schmitt, TUnfold: an algorithm for correcting migration effects in high energy physics. JINST **7**, T10003 (2012). arXiv: 1205.6201 [physics.data-an]
8. D. Calvetti, S. Morigi, L. Reichel, F. Sgallari, Tikhonov regularization and the l-curve for large discrete ill-posed problems. J. Comput. Appl. Math. **123**(1), 423–446 (2000). (Numerical Analysis 2000. Vol. III: Linear Algebra)
9. A. Nikolakopoulos, et al., Electron versus muon neutrino induced cross sections in charged current quasielastic processes. Phys. Rev. Lett. **123**, 052 501 (2019)
10. A.M. Ankowski, Effect of the charged-lepton's mass on the quasielastic neutrino cross sections. Phys. Rev. C **96**, 035 501 (2017)
11. T2K Collaboration, K. Abe, et al., Measurement of neutrino and antineutrino oscillations by the t2k experiment including a new additional sample of ν_e interactions at the far detector. Phys. Rev. D **96**, 092 006 (2017)

Chapter 5
Systematic Uncertainties for PRISM

In order to build predictions of the FD event rate in a PRISM oscillation measurement, it is necessary to accurately model the neutrino flux, neutrino interactions on nuclear targets and the near and far detectors. In particular, the PRISM method relies upon an accurate flux model to correctly combine the ND off-axis measurements. The models used in the analysis are subject to systematic uncertainties that affect the precision of the oscillation measurement. Systematic uncertainties are implemented as parameters of the models that, when varied, alter the components of the analysis that are derived from the MC simulation. This includes the ND and FD backgrounds, the FD flux-matching correction, the smearing matrices, the selection efficiencies and the linear combination coefficient calculation. Each of these analysis components are described in detail in Chap. 4. This chapter will discuss how the flux, cross section and detector systematic uncertainties enter the PRISM analysis and the impact they have on the PRISM oscillation measurement program.

The PRISM oscillation analysis is implemented in the CAFAna framework [1]. CAFAna includes the effect of systematic uncertainties by generating additional MC predictions that are shifted according to the particular systematic parameter. Typically, this will mean generating additional MC predictions shifted by ± 1, ± 2 and $\pm 3\,\sigma$ for each parameter. The 1σ in this context refers to the a priori uncertainty on the systematic parameter. A MC prediction can then be produced for an arbitrary set of systematic shifts for any number of parameters by performing a cubic interpolation between the different systematically shifted predictions.

5.1 Flux Uncertainties

Uncertainties in the neutrino flux primarily arise from uncertainties in the production of hadrons in the LBNF target and design uncertainties in the LBNF, such as the alignment of the graphite target and focusing horns. The neutrino flux prediction

C. M. Hasnip, *DUNE-PRISM—A New Method to Measure Neutrino Oscillations*, Springer Theses, https://doi.org/10.1007/978-3-032-07251-1_5

and the corresponding flux uncertainties are generated for the near and far detectors using the G4LBNF simulation and an adapted version of the Package to Predict the FluX (PPFX) framework, which was created for the MINERνA Collaboration [2, 3]. Flux uncertainties are generated for all off-axis positions at the ND. All of the MC components of the PRISM analysis, including the background predictions and the calculation of the linear combination coefficients, are affected by systematic uncertainties in the neutrino flux.

As discussed in Sect. 3.2, protons interact with the graphite target to produce secondary hadrons that subsequently decay into neutrinos. Although the most common hadrons produced are pions, other types of hadron such as kaons, which have different decay kinematics to pions, are also created in the target. The secondary hadrons may also interact with material in the beamline. There are uncertainties associated with modelling the production and subsequent interactions of hadrons in the beamline and these are collectively called the hadron production uncertainties. The accuracy of the hadron production model is significantly improved through the input of external measurements from fixed target experiments. The T2K Collaboration used measurements of pion [4] and kaon [5] production at the NA61/SHINE experiment to improve the T2K flux prediction [6]. Hadron production uncertainties are estimated using the PPFX framework. PPFX allows for the creation of many predictions of the LBNF flux by reweighting the nominal prediction with the hadron production model parameters thrown to random values. These many systematic "universes" quantify the systematic uncertainty in the parameters of hadron production model. In order to control the number of flux uncertainty parameters, the PPFX throws are combined into a total error matrix that describes the covariances between different neutrino energies, neutrino flavours and beam modes (i.e. FHC or RHC). This error matrix is then diagonalised to obtain the eigenvectors and eigenvalues. The eigenvectors are regarded as the "principle components" of the error matrix that quantify the variance described by the original error matrix. The principle components then become the hadron production uncertainty parameters propagated through the analysis. Smaller eigenvalues are associated with less important eigenvectors. Thus, only a subset of the most important eigenvectors with large eigenvalues are required to describe most of the variance due to the hadron production uncertainties. This technique is called principle component analysis (PCA) [7] and the hadron production uncertainties are described by the most important PCA parameters.

The second type of flux systematic uncertainties arise from the precision with which various components of the neutrino beam production facility are known. Although these are often broadly referred to as "focusing" uncertainties, this thesis breaks this broad group down into "focusing", "horn alignment" and "beam alignment" uncertainties. These systematic parameters are estimated by altering the parameters of the G4LBNF simulation. For example, to determine the impact of a shift in the horn current on the neutrino flux, the G4LBNF simulation is run with the horn current set at $+1\sigma$ relative to the nominal value. The resulting shifted flux is then compared to the nominal flux for each neutrino flavour and horn polarity. At the

Table 5.1 Uncertainties on the flux model parameters. The same set of systematic uncertainty parameters affects the flux at the ND and FD. Parameters are categorised based on the component of the flux model they impact

Parameter name	Category	1σ Uncertainty
PPFX	**Hadron Prod.**	100 PPFX throws
Decay pipe radius	**Focusing**	± 10 cm
Water layer		± 0.5 mm
Horn current		± 3 kA
Target density		± 1.8 %
POT counting		± 2 %
Horn 1 X-Shift	**Horn Alignment**	± 0.5 mm
Horn 2 X-Shift		± 0.5 mm
Horn 1 Y-Shift		± 0.5 mm
Horn 2 Y-Shift		± 0.5 mm
Beam θ	**Beam Alignment**	± 0.07 mrad tilt
Beam $\theta + \phi$		± 0.07 mrad tilt + 90° rotation
Beam σ		± 0.1 mm
Beam X-Offset		± 0.45 mm

ND, the flux uncertainties are calculated for every off-axis position and the shifted 280 kA on-axis uncertainties are assumed to be the same as for the nominal horn current.

Table 5.1 summarises all the flux systematic parameters, their categorisation and the corresponding uncertainty. The "Decay Pipe Radius" parameter has a large uncertainty of 10 cm in order to account for the many ways the geometry of the decay pipe could change other than simply a change in the radius. The "Water Layer" parameter accounts for the uncertainty in the thickness of the layer of cooling water that surrounds graphite target. Uncertainty in the density of the graphite target is quantified in the "Target Density" parameter. An overall normalisation uncertainty on the number of protons impacting on the target is captured by the "POT Counting" parameter. The names of the beam and horn alignment uncertainties in Table 5.1 are more self-explanatory. The horn alignment uncertainties describe the uncertainty in the position of the focusing horns relative to the incident proton beam in the transverse plane. However, for this set of uncertainties there is a change in naming convention for the focusing horns. As explained in Sect. 3.2, there are three focusing horns at the LBNF: Horns A, B and C. Changes in the Horn A position are expected to cause the largest change in the flux because of its close proximity to the target. Relatively small shifts in the position of Horn C are not expected to significantly effect the flux due to its distance from the target, hence Horn C alignment uncertainties are not considered. Only changes in the alignment of Horns A and B in the x and y directions are accounted for, creating a new set of horn alignment uncertainties for Horn 1 (Horn A) and Horn 2 (Horn B). The x direction in the ND target hall corresponds to movement left/right in the horizontal plane perpendicular to the beamline and the y direction refers to movement up/down in the vertical plane perpendicular to the beamline.

5.1.1 Flux Uncertainty Cancellation

The impact of the flux uncertainties on the PRISM analysis can be quantified by examining shifts in the ND and FD true CC event rate for each of the parameters set out in Table 5.1. Figure 5.1 shows the shift in the ND ν_μ CC event rate at each off-axis position due to a shift in the horn current. Flux uncertainty parameters may shift the flux at certain ND off-axis positions differently from the on-axis shift. This can be seen in Fig. 5.1, where a large 4% shift at approximately 4 GeV in the on-axis position becomes a smaller 1–2% shift at the off-axis positions. The shifts observed at off-axis positions tend to be at lower neutrino energies where the off-axis neutrino flux peaks.

The ND event rates at each off-axis position, shifted by the flux parameter of interest, are linearly combined to produce an oscillated prediction of the FD event rate. The linear combination coefficients are calculated using the nominal flux prediction. None of the corrections to account for backgrounds or selection efficiency and resolution differences between the ND and FD, as described in detail in Chap. 4, are required here as this study is only using true CC event rates. The fractional shift in the PRISM linear combination of true CC event rates can then be compared to the shift in FD MC prediction. Figure 5.2 shows the FD MC and PRISM linear combination predictions for $\nu_\mu \rightarrow \nu_\mu$ and $\nu_\mu \rightarrow \nu_e$ with a 1σ shift in the horn current parameter. The fractional shift in the predictions are also plotted at the bottom of each plot in Fig. 5.2. A different normalisation is chosen for the $\nu_\mu \rightarrow \nu_\mu$ and $\nu_\mu \rightarrow \nu_e$ channels. Shifts in the disappearance predictions are normalised by the unoscillated FD event rate, whilst shifts in the appearance predictions are normalised by the FD appearance

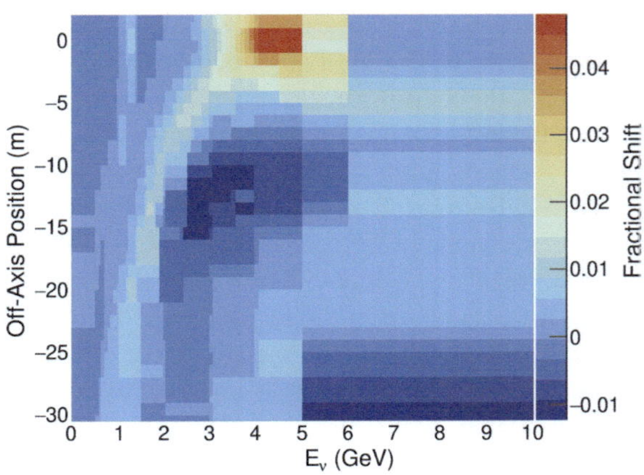

Fig. 5.1 Fractional change in the ND true ν_μ CC event rate due to a 1σ increase in the Horn Current parameter. In addition to the large shift on-axis at 4 GeV, there are smaller shifts in the flux at off-axis positions not present on-axis

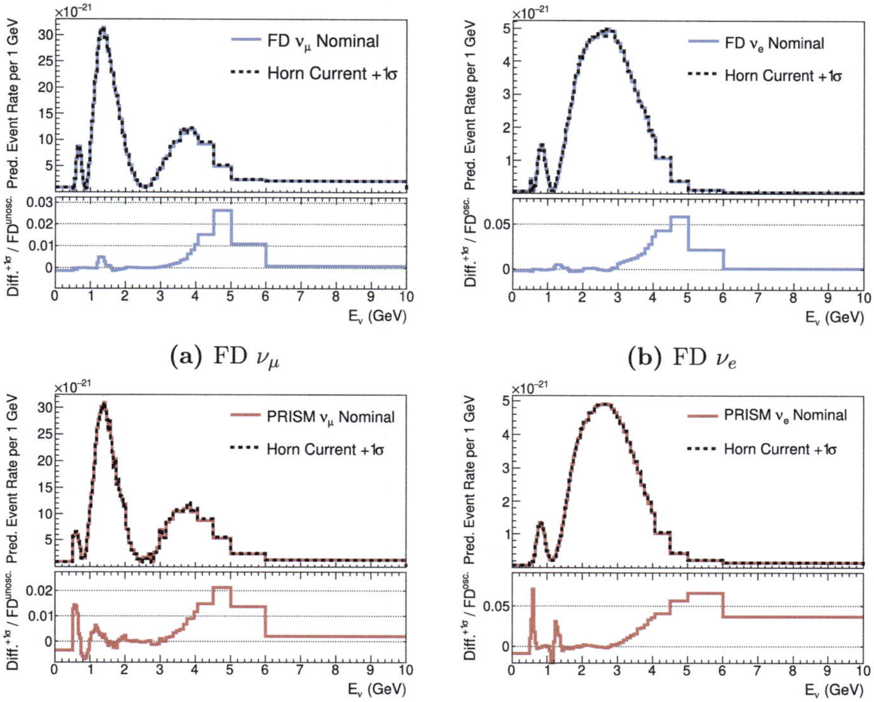

(a) FD ν_μ (b) FD ν_e

(c) PRISM ν_μ (d) PRISM ν_e

Fig. 5.2 Fractional change due to the Horn Current parameter in the FD MC oscillated event rate prediction and PRISM linear combination as a function of true neutrino energy. The PRISM prediction is generated by linearly combining the shifted ND event rates with coefficients calculated from the nominal MC. Similar fractional shifts can be seen in the FD MC and PRISM predictions in the 4 GeV region

event rate. This is because sensitivity to θ_{23}, which is dominated by the disappearance channels, comes from measuring the difference between the expected unoscillated flux and the observed ν_μ event rate. For the $\nu_\mu \rightarrow \nu_\mu$ predictions, notice how both the FD MC and PRISM predictions have a large shift due to the horn current parameter at approximately 4 GeV. However, the PRISM prediction has additional shifts of the order 1% at lower energies that are not present in the FD MC. This is due to the horn current parameter altering the neutrino flux at off-axis positions that are then added to the linear combination.

If the shifts in the near and far detector fluxes are similar, the effect of the flux uncertainty parameter on the analysis is expected to cancel. Total cancellation of the flux uncertainties does not occur due to neutrino oscillations changing the shape of the flux at the FD and the fact that the proximity of the ND to the beam production facility causes it to see a line source of neutrinos, whilst the FD sees a point source. Changes in the off-axis ND fluxes that are different from the change in the on-axis

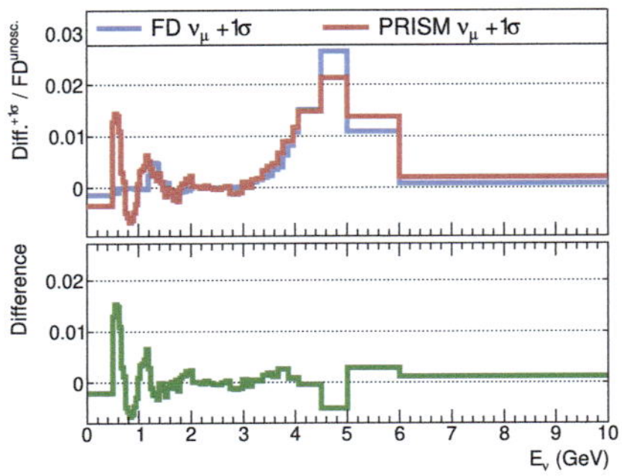

(a) ν_μ

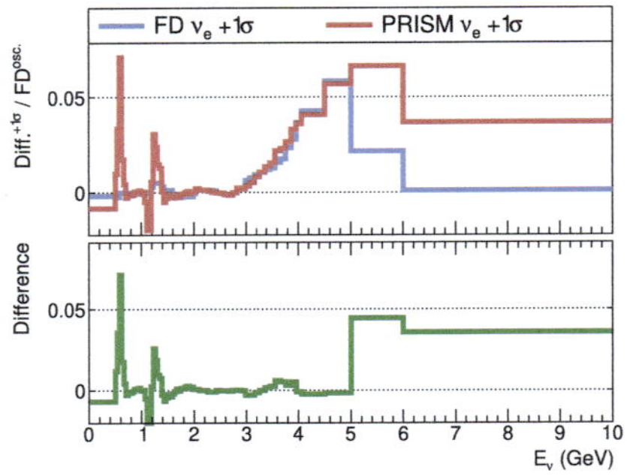

(b) ν_e

Fig. 5.3 The fractional shifts in the FD MC and PRISM linear combination predictions due to the Horn Current parameter seen in Fig. 5.2 are plotted together and the difference between them is taken. The difference between the shifts in the FD MC and PRISM predictions can be assumed to be the total systematic uncertainty in the linear combination. Good systematic cancellation is seen at 4 GeV

flux present an additional source of flux uncertainty in a PRISM analysis. The impact of flux systematic uncertainty cancellation is shown explicitly in Fig. 5.3. Systematic uncertainty in the linear combination of ND event rates due to a flux parameter can be regarded as the difference between the 1σ fractional shifts of the PRISM and FD MC predictions. Figure 5.3 demonstrates this cancellation in the 4 GeV region in both the disappearance (5.3a) and appearance (5.3b) channel. Both the PRISM and FD predictions exhibit a similar behaviour due to the change in horn current, leading to a systematic uncertainty close to zero above 2 GeV. However, the cancellation is not perfect at lower energies, thereby allowing the horn current parameter to still have an impact on the PRISM oscillation analysis through the linear combination.

The procedure followed in calculating the systematic uncertainty in the linear combination due to the horn current parameter shown in Fig. 5.3 is repeated for each of the flux systematic parameters set out in Table 5.1 and for all four neutrino flavour channels. Systematic uncertainty due to the hadron production uncertainties is quantified in Figs. 5.4 and 5.5. The first four PCA components of the hadron production uncertainties are shown separately in Fig. 5.4, as these are expected to be the largest effects. Figure 5.5 shows the uncertainties for the next four PCA components. Figures 5.6, 5.7 and 5.8 show the impact of the focusing, horn alignment and beam alignment parameters respectively. It is common across all of these studies to see large fractional uncertainties at particular low energy bins. This is particularly the case for the $\nu_\mu \rightarrow \nu_e$ and $\bar{\nu}_\mu \rightarrow \bar{\nu}_e$ channels, where across Figs. 5.4, 5.5, 5.6, 5.7 and 5.8 large fractional uncertainties are present at approximately 0.6 and 1.2 GeV. As can be seen from Fig. 5.2, these energies correspond to the minimum of the appearance oscillation probability for this choice of oscillation parameters, which exaggerates the size of the fractional shift calculated using the oscillated FD event rate

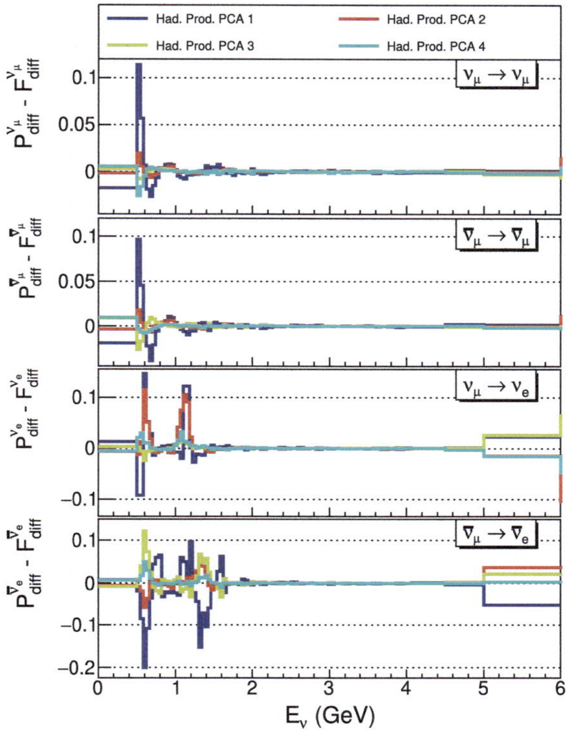

Fig. 5.4 Difference between the fractional shifts in the FD MC and PRISM linear combination due to hadron production uncertainty PCA parameters 1–4 for each oscillation signal channel. The symbols $P^{\nu_l}_{\mathrm{diff}}$ and $F^{\nu_l}_{\mathrm{diff}}$ refer to the fractional difference of the PRISM and FD MC prediction due to the systematic shift respectively. As expected, the first hadron production PCA component has the most significant effect. Reasonable cancellation occurs at higher energies

as the normalisation. A similar effect occurs for the $\nu_\mu \to \nu_\mu$ and $\bar{\nu}_\mu \to \bar{\nu}_\mu$ channels at 0.6 GeV, which is where the unoscillated FHC (RHC) ν_μ ($\bar{\nu}_\mu$) flux becomes small.

Above 1 GeV in the disappearance channels, Figures 5.4 and 5.5 demonstrate that there is reasonably good cancellation of the hadron production uncertainties between the near and far detectors. Cancellation of the hadron production uncertainties can also be seen in the appearance channels at energies removed from the appearance probability minima at 0.6 GeV and 1.2 GeV. The first four PCA components are expected to be the most important of the hadron production uncertainties. This can be seen when comparing Figs. 5.4 and 5.5, where PCA components 1–4 provide larger fractional uncertainties than components 5–8.

According to Figs. 5.7 and 5.8, two of the most significant horn and beam alignment systematic parameters "Horn 1 X-Position"[1] and "Beam X-Offset" respectively. The reason for this can be seen by looking at the change in the off-axis ND event rates. Figure 5.9 shows the change in the ND event rates due to the "Horn 1 X-Position" (5.9a) and "Beam X-Offset" (5.9b) parameters. In both cases, an approximately 1–

[1] The impact of the parameters "Horn 1 X-Position" and "Horn 2 X-Position" appears to be near-identical, which is not expected. Comparisons with a more recent set of flux systematic uncertainties shows that uncertainty in the positions of the secondary focusing horns has a much smaller impact on the event rate predictions than the first focusing horn, leading to the conclusion that "Horn 2 X-Position" is an inaccurate parameter that should be removed from the analysis.

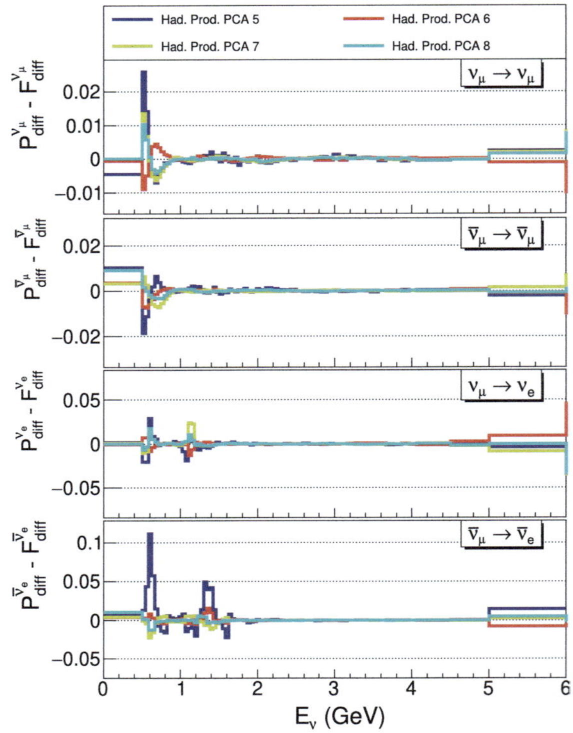

Fig. 5.5 Difference between the fractional shifts in the FD MC and PRISM linear combination due hadron production uncertainty PCA parameters 5–8 for each oscillation signal channel. The symbols $P^{\nu_l}_{diff}$ and $F^{\nu_l}_{diff}$ refer to the fractional difference of the PRISM and FD MC prediction due to the systematic shift respectively. The higher PCA components have smaller impact than the first 4 PCA components shown in Fig. 5.4

2% change in the event rates can be seen in the 1–2 GeV region off-axis, which is propagated through to the linear combination. These shifts in the ND off-axis flux do not cancel with FD shift, thereby contributing to the systematic uncertainty in the PRISM prediction.

5.1.2 Flux Uncertainties in a PRISM Measurement

The studies shown in Figs. 5.4, 5.6, 5.7 and 5.8 use the nominal MC to calculate the linear combination coefficients. Systematic shifts are applied to the ND event rates, which are linearly combined to make a PRISM prediction, and the FD MC. This is done to explicitly show the degree of flux systematic cancellation between the near and far detectors in a PRISM oscillation analysis. However, in a realistic PRISM analysis measured data at the ND is linearly combined to predict FD data. Data from the detectors are not varied by systematic uncertainties. As stated at the beginning of this chapter, in a PRISM oscillation analysis the MC components are varied by systematic uncertainties. This includes corrections for backgrounds, selection efficiency and resolution, as well as the calculation of the linear combination coefficients.

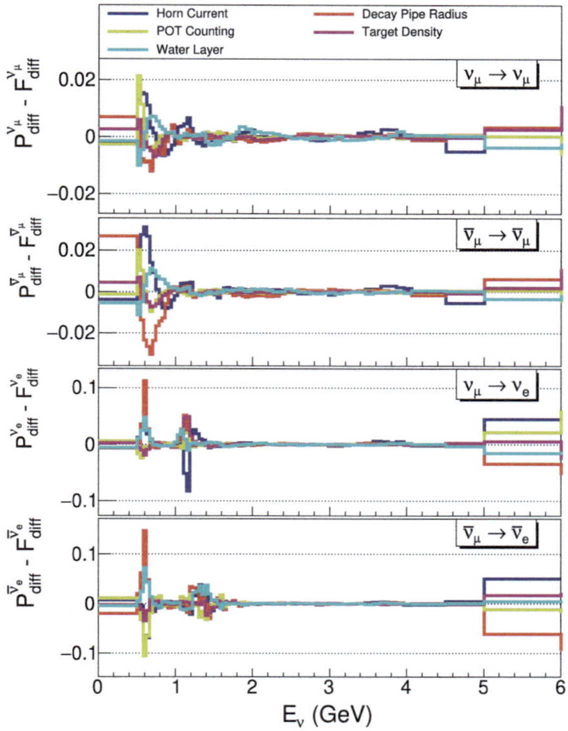

Fig. 5.6 Difference between the fractional shifts in the FD MC and PRISM linear combination due to the focusing uncertainty parameters for each oscillation signal channel. The symbols $P^{\nu_l}_{diff}$ and $F^{\nu_l}_{diff}$ refer to the fractional difference of the PRISM and FD MC prediction due to the systematic shift respectively. The uncertainty in the decay pipe radius has a substantial impact due to the very large a priori uncertainty assigned to this parameter

Cancellation of flux systematic uncertainties in a realistic PRISM analysis primarily occurs in the calculation of the linear combination coefficients. The linear combination coefficients are calculated by mapping off-axis ND CC event rates to the FD CC event rate. Section 4.4 describes this calculation in detail. If the ND event rates are shifted in a way that is very different from the FD event rate, then there will be a large change in the calculated linear combination coefficients. These shifted coefficients are then applied to the ND data in a linear combination, resulting a shifted PRISM prediction. Likewise, if the nature of ND and FD event rate shifts are very similar, then the change in the coefficients will be small, resulting in a PRISM prediction little different from the nominal prediction.

The impact of the systematic uncertainties on the PRISM prediction through the variation of all the MC components can be studied by building covariance matrices for the systematic uncertainty. Many random throws of the selected group of systematic parameters are performed and the covariance between each histogram bin of the shifted and nominal PRISM prediction is calculated. A 1σ error band can be extracted by taking the square-root of the variance of each PRISM prediction histogram bin, which corresponds to the diagonal elements of the covariance matrix. Caution should be taken in interpreting the impact of the systematic parameters based on a 1σ error band that ignores covariances between the bins of the predictions. However, such

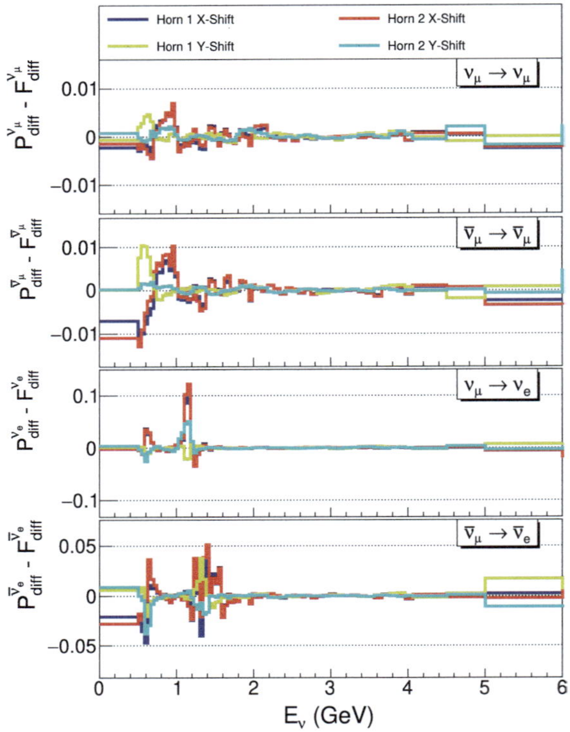

Fig. 5.7 Difference between the fractional shifts in the FD MC and PRISM linear combination due to the horn alignment uncertainty parameters for each oscillation signal channel. The symbols $P^{\nu_l}_{diff}$ and $F^{\nu_l}_{diff}$ refer to the fractional difference of the PRISM and FD MC prediction due to the systematic shift respectively. The parameter Horn 1 X-Shift has an important effect due to changes in the flux off-axis that do not cancel with changes in the FD flux

studies provide insight into the expected importance of different systematic parameters. The 1σ error band quantifies the expected variation of the PRISM prediction in an oscillation analysis due to a set of systematic parameters. It is the relative systematic error on the predicted event rate.

Flux systematic uncertainties impact the PRISM analysis through the calculation of the linear combination coefficients and the corrections for ND and FD backgrounds, selection efficiencies and resolutions. The dominant MC component through which the flux uncertainties alter the PRISM prediction depends on the neutrino flavour channel. Figure 5.10 plots 1σ error bands extracted from the covariance matrices produced by varying all flux systematic parameters. There are two 1σ error bands in Fig. 5.10, one where all the MC components of the PRISM prediction are allowed to be varied by the flux systematic uncertainties and one where the calculation of the linear combination coefficients is not allowed to vary. It can be seen in the ν_μ window of Fig. 5.10 that most of the variation in the PRISM prediction is due to the flux uncertainties altering the calculation of the linear combination coefficients. However, for the RHC $\bar{\nu}_\mu$ and $\bar{\nu}_e$ predictions, allowing the coefficient calculation to varied by the flux systematic uncertainties adds little to the 1σ error band, meaning most of the variation due to the flux uncertainties enters due to the MC backgrounds

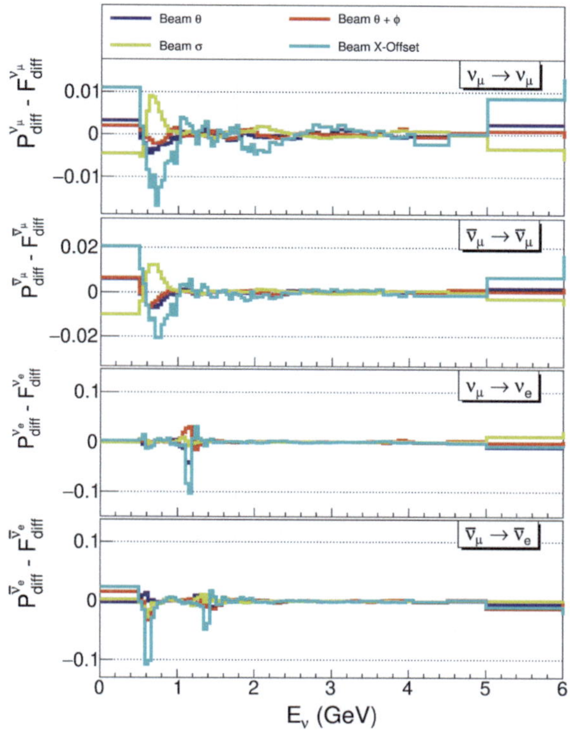

Fig. 5.8 Difference between the fractional shifts in the FD MC and PRISM linear combination due to the beam alignment uncertainty parameters for each oscillation signal channel. The symbols $P^{\nu_l}_{diff}$ and $F^{\nu_l}_{diff}$ refer to the fractional difference of the PRISM and FD MC prediction due to the systematic shift respectively. The parameter Beam X-Offset has an important effect

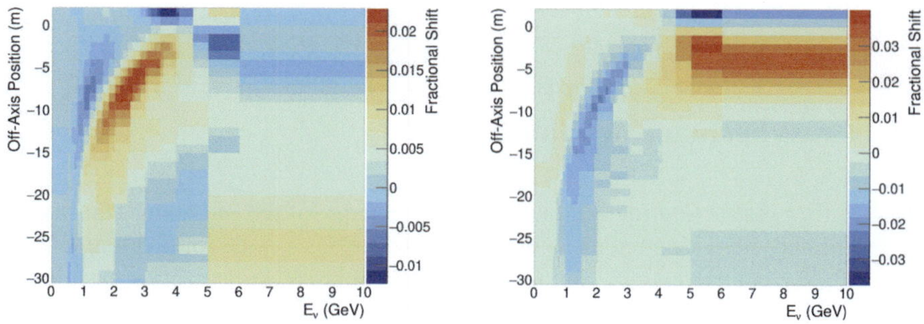

(a) Horn 1 X-Shift (b) Beam X-Offset

Fig. 5.9 Fractional change in the ND true ν_μ CC event rate due to a 1σ increase in the Horn 1 X-Shift and Beam X-Offset parameters. The shifts observed at off-axis position in the ND fluxes for these parameter will not necessarily cancel with any change in the FD flux, leading to a larger contribution to the systematic uncertainty

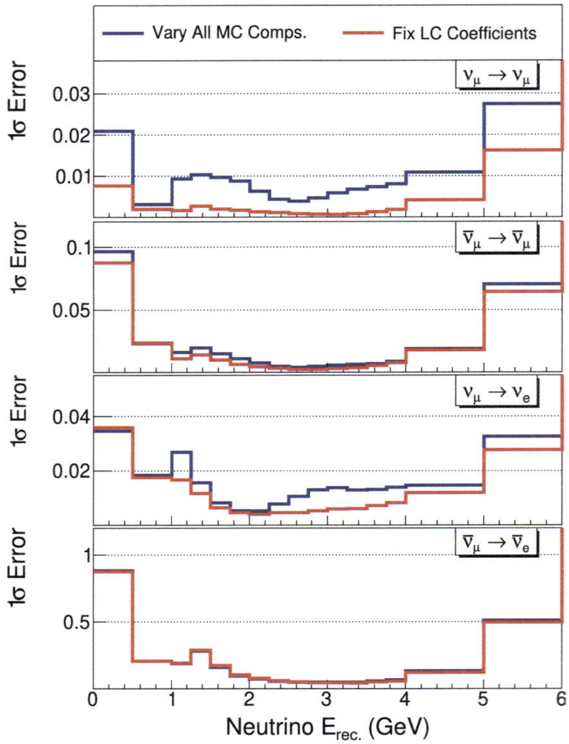

Fig. 5.10 Comparison of the 1σ error band when all the MC components of the total PRISM prediction are able to vary and when the linear combination coefficient calculation is fixed. The magnitude of the 1σ systematic uncertainty band is plotted as a function of reconstructed neutrino energy for each oscillation channel. The magnitude of the 1σ error band for the ν_μ prediction reduces substantially when the linear combination coefficient calculation is fixed

and flux-matching correction. This is expected for RHC mode as the backgrounds, particularly the wrong-sign background, form a larger component of the total PRISM prediction than the FHC mode backgrounds.

5.2 Cross Section Uncertainties

The neutrino interaction model relates the true energy of the incoming neutrino to the multiplicity and kinematics of the final state particles in an interaction. Uncertainties in the neutrino interaction model are referred to as cross section systematic uncertainties. The default neutrino interaction model and the associated uncertainties are contained in the 2.12.10 version of GENIE used to generate the MC events [8]. Uncertainties in the neutrino interaction model are implemented by varying parameters in the GENIE generator, through additional weights that parameterise cross section effects not modelled in GENIE and by comparing GENIE with other neutrino interaction models [9]. A list of the systematic parameters modelled in GENIE is presented in Table 5.2. Parameters modelling additional effects not included in

Table 5.2 Uncertainties on the interaction model parameters in the GENIE generator. The variable W refers to the energy transferred to the nucleus in a neutrino interaction. Events with a low-W that are not CCQE tend to be resonance production events. High-W events tend to be DIS. The initials "BY" refer to the Bodek-Yang model of DIS [10]

Parameter name	Category	1σ Uncertainty
Axial mass for CCQE	**Quasielastic**	$^{+0.25}_{-0.15}$ GeV
CCQE vector form factors		BBA05 or Dipole
Fermi surface momentum		±30% GeV
Axial mass for CC resonance	**Low-W**	±0.05 GeV
Vector mass for CC resonance		±10%
θ_π distribution for Δ decay		N/A
A_{HT} in BY model	**High-W**	±25%
B_{HT} in BY model		±25%
C_{V1u} correction in BY model		±30%
C_{V2u} correction in BY model		±40%
Axial mass for NC resonance	**Other NC**	±10%
Vector mass for NC resonance		±5%
Nucleon charge exchange	**Intra-nuclear**	±50%
Nucleon elastic reaction		±30%
Nucleon inelastic reaction		±40%
Nucleon absorption		±20%
Nucleon π-production		±20%
π charge exchange		±50%
π elastic reaction		±10%
π inelastic reaction		±40%
π absorption		±20%
π π production		±20%

GENIE are listed in Table 5.3. Uncertainties in the neutrino interaction model affect all the MC components of the analysis except the linear combination coefficients.

The nucleons inside the argon nucleus occupy some initial state before the interaction with an incoming neutrino. GENIE models this initial state of nucleons in the nucleus with a modified version of the Bodek and Richie global Fermi gas model [11]. GENIE attempts to improve the accuracy of the Fermi gas model by including empirical corrections for short-range correlations between nucleons, which populates the high-momentum tails of the nucleon momentum distributions in the Fermi gas model.

Cross section uncertainties that affect the modelling of CCQE interactions are dominated by the axial form factor of the nucleus and random phase approximation (RPA) calculations. The model for the RPA calculations comes from Nieves et al. [12]. The uncertainty in the axial form factor is quantified by the axial mass parameter M_A. The M_A parameter tends to over-constrain the axial form factor at high momentum transfers (Q^2). To correct for this an RPA effect parameterised by

Table 5.3 Additional cross section uncertainty parameters not contained in GENIE. There are three BeRPA parameters covering different regions of Q^2 space and 23 DIS parameters for CC/NC neutrino/antineutrino scattering on protons/neutrons with one to three pions in the final state. Separate 2p-2h parameters are implemented for neutrino and antineutrino scattering

Parameter name	Mode	Description
BeRPA	1p-1h/QE	RPA/nuclear model suppression
ArC2p2h	2p-2h Ar/C	Electron scattering SRC pairs
E_{2p2h}	2p-2h	2p-2h Energy dependence
CC Non-resonant	ν DIS	$\nu + n/p \to l + 1\pi$
Other Non-resonant	$N\pi$ DIS	$1 < W < 5$ GeV
ν_e/ν_μ	ν_e, $\bar{\nu}_e$	Cross section ratio
$\nu_e/\bar{\nu}_e$	ν_e, $\bar{\nu}_e$	Cross section ratio

a Bernstein polynomial is included (BeRPA). There are several BeRPA parameters controlling the shape of the Bernstein polynomial at different values of Q^2. These BeRPA parameters were developed by the T2K Collaboration to complement the existing GENIE CCQE uncertainty parameters [13].

The Nieves model of 2p-2h interactions (see Sect. 2.6 for details) is incorporated into the GENIE cross section calculation. It was demonstrated in MINERνA that the standard Nieves model under-predicted the strength of 2p-2h interactions on carbon targets and there is very limited data available for scattering on argon [14]. Systematic parameters are therefore added that account for the uncertainty associated with the strength of the 2p-2h interaction and the scaling of the Nieves model prediction from a carbon to an argon target. The uncertainties on the scaling of 2p-2h interactions from carbon to argon are taken from electron scattering measurements on short-range correlated (SRC) nucleon pairs [15]. Both of these 2p-2h parameters are listed in Table 5.3.

The neutrino events generated by GENIE use the Rein and Sehgal model for single pion production in resonance events [16]. This model for single pion resonance events is tuned to deuterium-scattering bubble chamber data [17]. Systematic parameters listed in Table 5.2 quantify the uncertainty in the axial and vector masses for CC resonance production.

Deficiencies in the default GENIE implementation of the transition region between resonant and deep inelastic scattering (DIS) events were found by the NOνA Collaboration [18]. This lead the DUNE Collaboration to develop a set of custom systematic parameters, listed in Table 5.3, that describe the uncertainty in modelling CC and NC DIS neutrino and antineutrino interactions on neutrons and protons with one or more pions in the final state. This corresponds to 23 systematic parameters that quantify the uncertainty in the neutrino interaction model in the resonant scattering and DIS transition region.

An intra-nuclear cascade model for final state interactions of hadrons in the nucleus is implemented in GENIE. This is called the *hA* model. GENIE provides a range of systematic parameters to quantify the uncertainty in the model predictions of final state interactions, which are listed in Table 5.2.

Neutrino cross sections in CCQE interactions are altered by the lepton flavour in the final state. The different final state lepton masses resulting from a ν_μ or ν_e interaction alters the cross section by changing the final state kinematic limit and the effect of radiative corrections. The neutrino interacts with a quark bound in a nucleon, therefore the cross section calculation accounts for the nucleon form factor. Uncertainties arise from differences in the nucleon form factor calculation due to the different final state lepton flavour [19]. Systematic parameters are implemented that account for the uncertainty in the ratios $\sigma_{\nu_e}/\sigma_{\nu_\mu}$ and $\sigma_{\nu_e}/\sigma_{\bar{\nu}_e}$, which are listed in Table 5.3.

5.2.1 Cross Section Parameter Impact

As in Sect. 5.1.2, the impact the numerous cross section parameters have on the PRISM analysis can be quantified by building covariance matrices for groups of cross section systematic parameters. The resulting 1σ error band can quantify the variation of the PRISM prediction due the cross section systematic uncertainties. Figure 5.11 groups the cross section parameters into the GENIE categories shown in Table 5.2 and plots the 1σ error band for each category of cross section parameters and each oscillation channel. As stated in Sect. 5.1.2, the 1σ error band should only be used as an indication of the most important systematic effects and not to quantify the total effect of a set of systematic parameters in an oscillation analysis.

According to Fig. 5.11, the CCQE parameters have little affect on the PRISM disappearance prediction at energies above 2 GeV in reconstructed neutrino energy. This is understandable, since there are few CCQE interactions at higher neutrino energies. Events with a high energy transfer to the nucleus ("High W") are likely to be DIS interactions, therefore the "High W" parameters become most important at high neutrino energies. Figure 5.11 also shows the "Other NC" parameters have the highest impact below 3 GeV in reconstructed energy, which is where majority of the NC backgrounds events are found.

Figure 5.12 shows the size of the 1σ error band for the different types of non-GENIE cross section uncertainties set-out in Table 5.3. The non-resonant DIS parameters only have an effect at high reconstructed neutrino energies and this is consistent across all four signal channels. This is expected, since DIS interactions generally occur at higher neutrino energies. The cross section ratio uncertainty parameters have practically no effect on the disappearance channels, but are one of the dominant contributions to the uncertainty in the appearance channels at low neutrino energies. Both the RPA effects (controlled by the BeRPA parameters) and 2p-2h interactions are most relevant for CCQE-like interactions, causing their impact on the 1σ error band to be small above approximately 2 GeV in reconstructed energy.

It can be interesting to investigate how the cross section ratio uncertainties have a large impact on the PRISM appearance channels at low energies. A likely cause is the

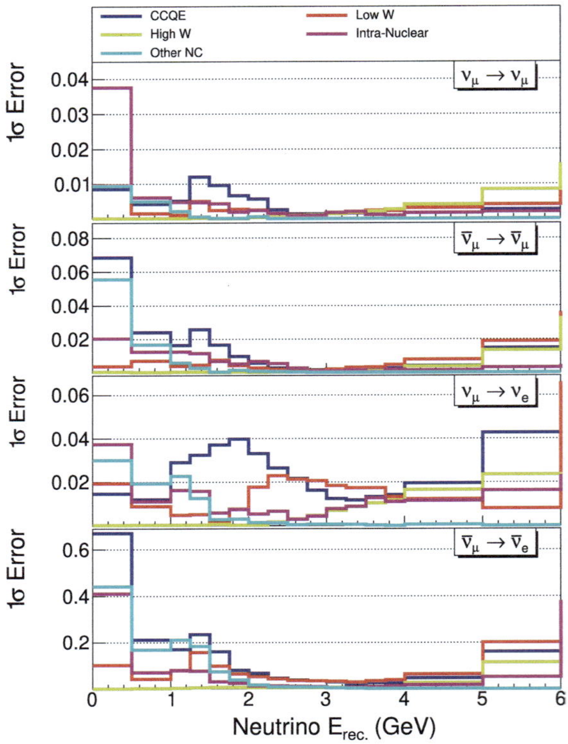

Fig. 5.11 The magnitude of the 1σ systematic uncertainty band is plotted as a function of reconstructed neutrino energy for each oscillation channel and each category of GENIE cross section systematic uncertainties set out in Table 5.2. Each error band is the combined effect of all the parameters associated with each category of cross section uncertainties set out in Table 5.2

reliance on the MC to calculate a cross section correction for the PRISM appearance prediction. More details on this can be found in Sect. 4.3.3. To test this, 1σ error bands were made when varying the PRISM predictions by the cross section ratio parameters in Table 5.3 for two cases. Firstly, when allowing all MC components of the PRISM prediction to be varied by the systematic parameters as usual. The second, when the MC predictions used to calculate the cross section correction are not allowed to be varied by systematic uncertainties. Figure 5.13 shows the result of this study, where the impact of the cross section ratio parameters is substantially reduced in the case where the cross section correction is not allowed to vary. This is especially true for the FHC $\nu_\mu \to \nu_e$ prediction.

5.2.2 Near Detector Efficiency Calculation

The MC component through which the cross section systematic uncertainties have the largest impact is the calculation of the ND selection efficiency. More details on this component of the PRISM analysis can found in Sect. 4.3.2. The effect of the cross section systematic uncertainties on the ND selection efficiency calcula-

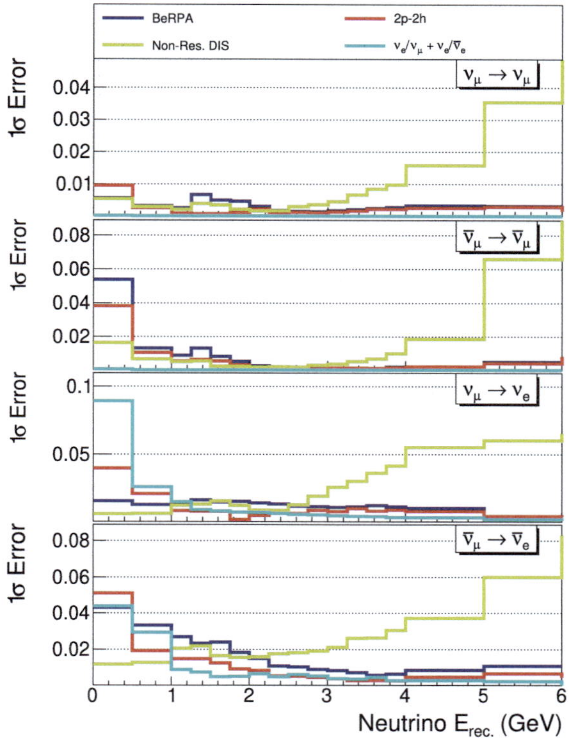

Fig. 5.12 The magnitude of the 1σ systematic uncertainty band is plotted as a function of reconstructed neutrino energy for each oscillation channel and each type of non-GENIE cross section systematic uncertainty set out in Table 5.3. The legend refers to the effect of the different cross section uncertainty types in Table 5.3. For example, the "Non-Res. DIS" error band includes the effect of all the non-resonant CC and non-CC DIS parameters

tion can be investigated by again building covariance matrices and plotting the 1σ band for the relative systematic error on the PRISM prediction. For one error band all the MC components are allowed to be varied by the cross section parameters, but for the second error band the calculation of the ND selection efficiency calculation is fixed, thereby removing the effect of this MC component on the error band. Figure 5.14 demonstrates how the ν_μ and ν_e PRISM predictions see substantial reductions in the magnitude of the 1σ error band in the 1–5 GeV region when the ND efficiency correction is not allowed to be varied by the cross section systematic parameters. However, a smaller change is seen in the $\bar{\nu}_\mu$ and $\bar{\nu}_e$ PRISM predictions and this is due to the much larger background component in RHC mode.

Figure 5.10a and b in Chap. 4 show the ND efficiency to be as low as 20% at some energies and off-axis positions. This means there are large differences between the distribution of selected ND events and the true ND signal events. Consequently, shifts in the selected and true signal ND events due to changes in the cross section parameters do not cancel well when the ratio of these event rates is taken to calculate the efficiency (Eq. 4.4). Correcting for such low ND selection efficiencies with the MC thus makes this component of the analysis highly sensitive to the cross section systematic uncertainties.

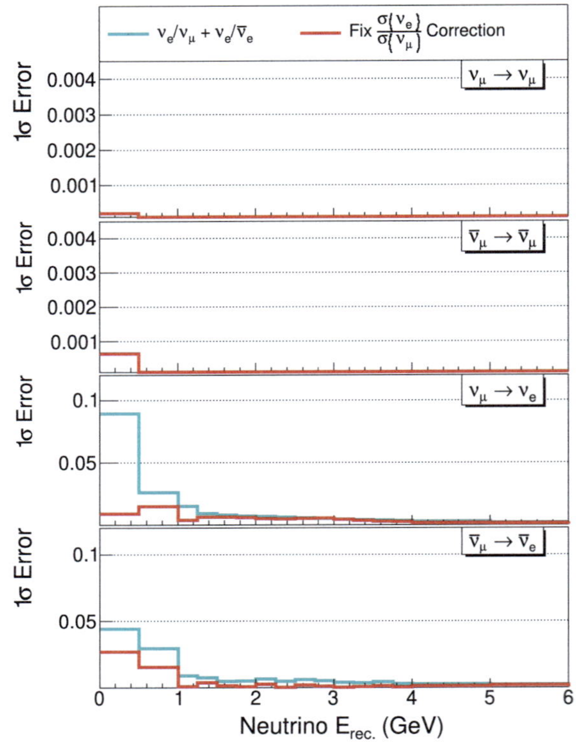

Fig. 5.13 The magnitude of the 1σ systematic uncertainty band is plotted as a function of reconstructed neutrino energy for each oscillation channel and for the cross section ratio uncertainties contained in Table 5.3. The cyan line is the same as in Fig. 5.12. However, for the red error band the calculation of the cross section ratio correction for the PRISM appearance channel (see Sect. 4.3.3 for details) is fixed. This substantially reduces the impact of the cross section ratio parameters

5.3 Detector Uncertainties

Uncertainties in the detector model alter the relationship between the true and reconstructed energy of the final state particles in the detector as well the event selection performance and background predictions. However, due to the parameterised reconstruction and simple event selection in the ND, this analysis limits itself to detector energy scale and resolution uncertainties. Energy scale and resolution uncertainties are implemented in the near and far detectors. Some of the energy scale uncertainties are wholly correlated between the ND and FD, whilst others are uncorrelated. Resolution uncertainties for the individual particle interactions are conservatively assumed to be totally uncorrelated.

A total energy scale is assigned to the ND and FD that is assumed to be totally uncorrelated due to significant differences in the calibration strategy for the two detector facilities. Energy scale and resolution uncertainties are then assigned to each type of particle response in the detectors. The individual energy scales of the muons, charged hadrons (protons and $\pi^{\pm}$) and EM showers (electrons and π^0) are correlated between the ND and FD because both near and far detector facilities measure particle energy deposits in liquid argon. This would be the case for the neutron energy scale uncertainty as well. However, as shown in Eq. 4.2, the neutron energy

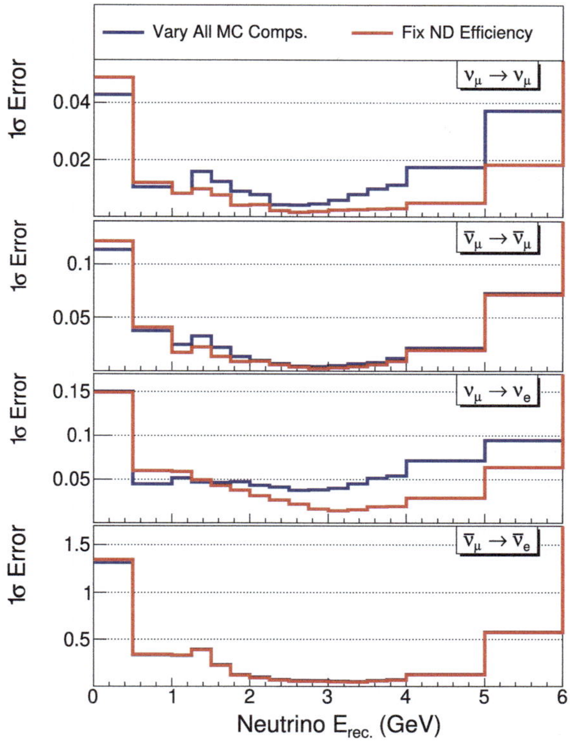

Fig. 5.14 Comparison of the 1σ error band when all the MC components of the total PRISM prediction are able to vary and when the ND efficiency calculation is fixed. The magnitude of the 1σ systematic uncertainty band is plotted as a function of reconstructed neutrino energy for each oscillation channel. The magnitude of the 1σ error band in generally reduces substantially when the ND efficiency correction is fixed

is not included in the ND analysis variable. This is because the full event reconstruction is only available in the FD MC, which is capable of reconstructing some energy from neutrons scattering on argon, whilst the parameterised ND reconstruction would likely overestimate the neutron energy deposits visible in the ND. As stated in Sect. 4.1.1, the neutron energy is also omitted from the ND analysis variable in order to unfold the ND effects using only quantities that are observable in the ND. Assuming the muon energy scale uncertainties are fully correlated between the ND and FD is too simplistic, since a large fraction of the muons at the ND have their energy measured in the TMS/ND-GAr downstream tracker rather than by range in liquid argon. TMS measures the muon energy by range in an iron/scintillator plane tracker and ND-GAr, which is the assumed downstream tracker for the available ND MC, measures the muon momentum by the curvature of the track. The energy scale of neither of these detectors is expected to be well-correlated with the FD module and this will be corrected in the future.

The uncertainty in the resolution of the detector response to a specific particle type is calculated from the difference between the reconstructed and true particle energy. The resolution uncertainty parameter, p_{res}, is assumed to be 2% for muons, charged hadrons and EM showers and 10% for neutrons. There are separate uncorrelated resolution uncertainties for the ND and FD. The neutron resolution uncertainty is only relevant to the FD since neutron energy is not included in the ND analysis variable. The reconstructed energy of a particle is varied by the resolution uncertainty parameter in the following way:

$$E'_{rec} = E_{rec} + p_{res}\left(E_{true} - E_{rec}\right), \tag{5.1}$$

where E'_{rec} and E_{rec} are the shifted and nominal reconstructed energy respectively and E_{true} is the true energy of the particle.

The energy scale uncertainties are allowed to vary with energy according to three free parameters—p_0, p_1 and p_2—with a priori constraints such that

$$E'_{rec} = E_{rec}\left(1 + p_0 + p_1\sqrt{E_{rec}} + \frac{p_2}{\sqrt{E_{rec}}}\right), \tag{5.2}$$

where E'_{rec} and E_{rec} are again the shifted and nominal energy respectively. The expected size of the energy scale and resolution uncertainties is derived from the experience of past neutrino experiments. For example, NOνA achieved an uncertainty on the energy scale of muon and protons of 1 and 5% respectively [18]. The total energy scales of the ND and FD are controlled by calibrating the detectors. Reconstructing the energy of neutrons remains a significant challenge, therefore a large uncertainty is assumed for the neutron energy scale in the FD. In the ND, a distinction is made between muons that are reconstructed by range or by curvature, which corresponds to a muon being contained in ND-LAr or propagating to ND-GAr respectively. Details of all the energy scale uncertainties are shown in Table 5.4.

The functional form of the energy scale uncertainties shown in Table 5.4 has been plotted to illustrate the variation in the energy scale uncertainty as a function of energy. Figure 5.15 shows how p_0, p_1 and p_2 each contribute to the overall energy scale uncertainty for individual particle responses and the total detector. The general pattern is for the p_2 parameter to cause the energy scale uncertainty to increase rapidly at very low energies and the p_1 parameter to slowly raise the uncertainty at

Table 5.4 Uncertainties on the energy scale parameters. The same parameters are used for the ND and FD energy scale uncertainties. Correlated uncertainties are 100% correlated between the ND and FD. Uncorrelated uncertainties are totally uncorrelated

Particle type	Correlated?	p_0 (%)	p_1 (%)	p_2 (%)
All (except muons)	No	2	1	2
μ (range)	Yes	2	0.5	2
μ (curvature)	Yes	1	0.5	1
$p, \pi^{\pm}$	Yes	5	5	5
e, γ, π^0	Yes	2.5	2.5	2.5
n	No	20	30	30

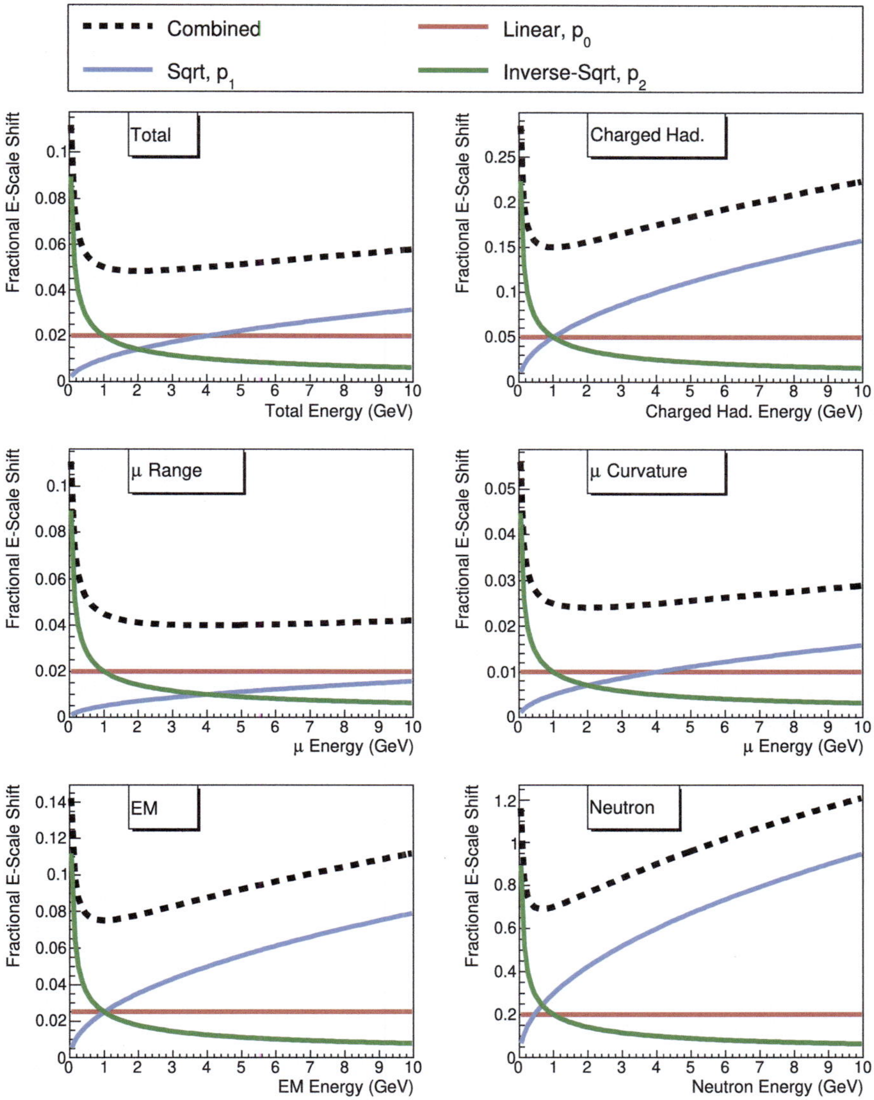

Fig. 5.15 Functional form of the energy scale uncertainty parameters p_0, p_1 and p_2 for each particle interaction type and the total energy scale. The combined effect is the sum of the p_0, p_1 and p_2 parameters

very high energies. In the case of the muon energy scale uncertainty, the p_1 parameter is suppressed as the energy scale is not expected to become less well-known as the length of the muon track increases.

The magnitude of a series of 1σ error bands that quantify the variation of the PRISM prediction due to the detector uncertainties are plotted. Figure 5.16 shows the 1σ error bands for the uncorrelated total energy scale uncertainties in the ND and FD. Both the ND and FD error bands vary the p_0, p_1 and p_2 parameters in Table 5.4 simultaneously. The size of the ND total energy scale error band is significantly smaller than the FD error band. This may be due to the use parameterised reconstruction in the ND MC. As can be seen when comparing the ND and FD smearing matrices in Figs. 4.10 and 4.13 respectively, the parameterised reconstruction provides much narrower resolutions than the realistic FD reconstruction. The detector uncertainties alter the relationship between the true and reconstructed energies of particles observed in the detector, hence the very diagonal ND smearing matrix will not be affected by the detector uncertainties as much as the FD smearing matrix.

The 1σ error bands are then plotted for the detector response uncertainties for different particle types in Fig. 5.17. The energy scale uncertainties for the individual detector responses are fully correlated between the ND and FD, except in the case of the neutrons. As in Fig. 5.16, each 1σ error band corresponds to the combined effect of

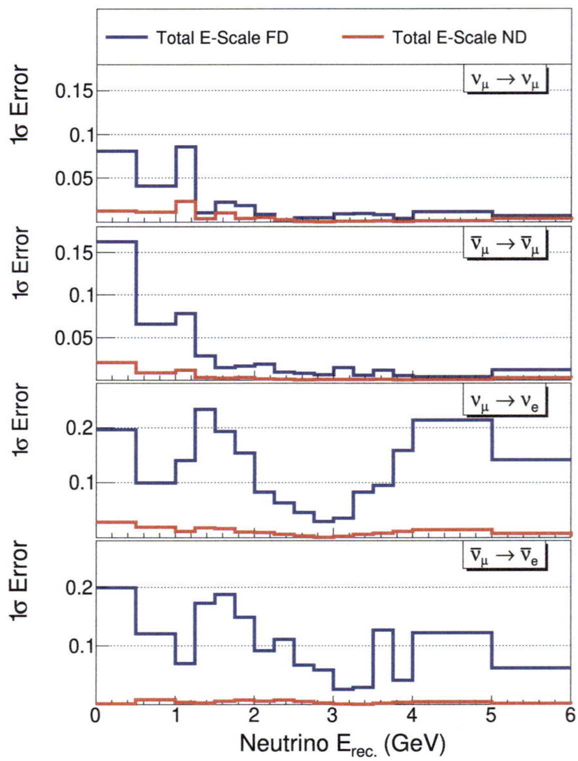

Fig. 5.16 Comparison of the 1σ error bands when varying the PRISM prediction by the p_0, p_1 and p_2 parameters for total energy scale uncertainty in the ND (red) and FD (blue)

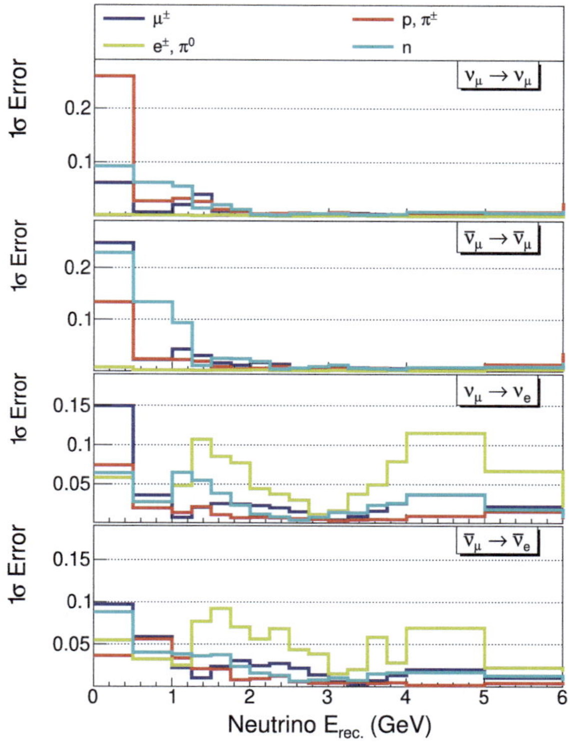

Fig. 5.17 Comparison of the 1σ error band when varying the PRISM prediction by the individual detector response uncertainties. For example, the $\mu^\pm$ error band quantifies the collective impact of the p_0, p_1 and p_2 parameters for the muon energy scale uncertainty that is correlated between the ND and FD

the p_0, p_1 and p_2 parameters for each response type. In the case of the disappearance predictions, the charged hadron and neutron energy scale uncertainties have the largest impact on the prediction, however the effect is concentrated below 1 GeV. As might be expected, the dominant uncertainty above 1 GeV in the appearance channels is the EM shower energy scale.

Resolution uncertainties quantify the uncertainty on the resolution of the detector reconstruction to individual interaction types. Figure 5.18 shows two error bands for the ND and FD resolution uncertainties. All of the ND resolution uncertainties are varied collectively to create the ND resolution uncertainty error band and likewise for the FD error band. The FD resolution uncertainties have an approximately 10% effect at energies below 1 GeV. However, the ND resolution uncertainties have practically no effect on the PRISM prediction. Like in the case of the total ND energy scale uncertainties (Fig. 5.16), this could also be due to the narrow resolution of the parameterised reconstruction used in the ND MC.

Fig. 5.18 Comparison of the 1σ error band when varying the PRISM prediction by the resolution uncertainties in the ND (red) and FD (blue). Each error band is made by collectively varying all the resolution uncertainties in the corresponding detector

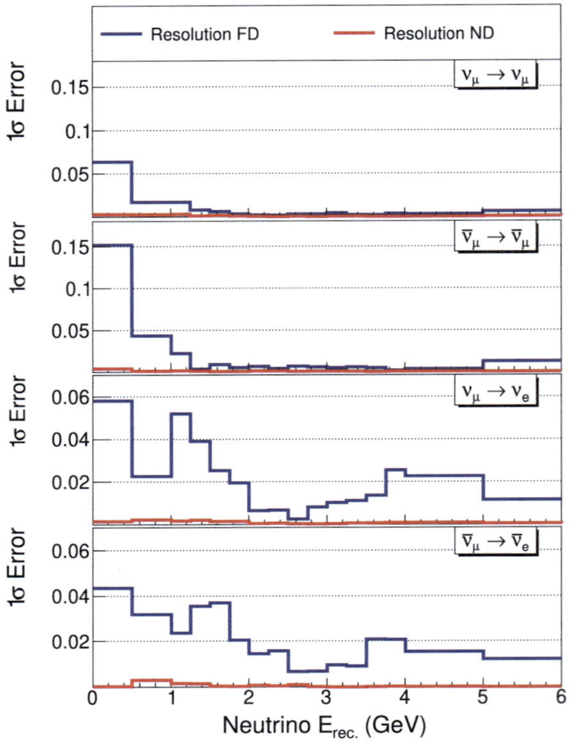

Finally, the primary means by which the detector uncertainties enter the PRISM analysis is studied in Fig. 5.19. The 1σ error bands are plotted for the combined effect of all the energy scale and resolution uncertainties in the ND and FD. In one error band the ND and FD smearing matrices are not allowed to be varied by the detector uncertainties. This drastically reduces the relative detector systematic error on the PRISM prediction. Hence, it is concluded that the method for correcting for resolution differences between the ND and FD via the smearing matrices is the dominant way the detector uncertainties enter the PRISM analysis. This is to be expected; the energy scale and resolution uncertainties are designed to alter the relationship between the true and reconstructed energy of the final state particles and this relationship is defined by the ND and FD smearing matrices.

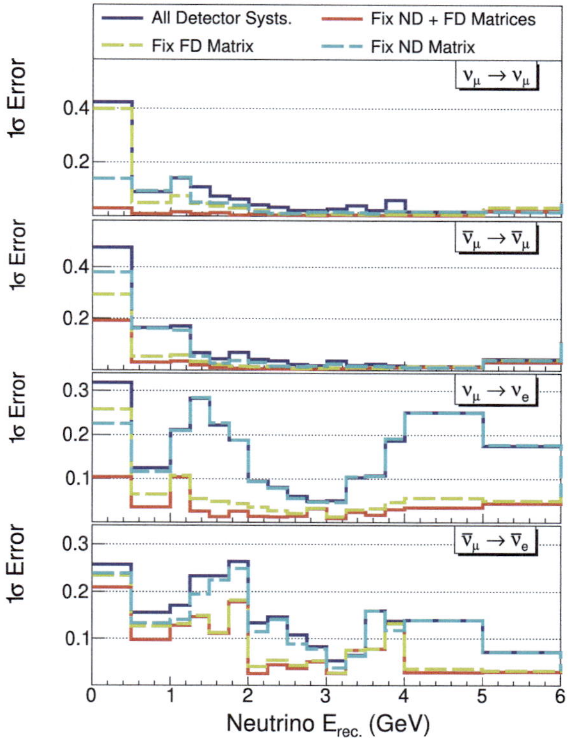

Fig. 5.19 Comparison of the 1σ error band when varying the PRISM prediction by all the detector uncertainties simultaneously. The two solid lines correspond to error bands where all the MC components are allowed to be varied by the detector systematic parameters (blue) and where the ND and FD smearing matrices are prevented from varying in the fit (red). The two dotted lines are when either the FD or ND smearing matrix is fixed. The size of the error band dramatically reduces when the smearing matrices are fixed

References

1. NOνA Collaboration, Nova-art, ch. cafana overview (2019), https://cdcvs.fnal.gov/redmine/projects/novaart/wiki/CAFAna_overview
2. MINERvA Collaboration, L. Aliaga, et al., Neutrino flux predictions for the numi beam. Phys. Rev. D **94**, 092 005 (2016)
3. T. Golan, L. Aliaga, M. Kordosky, Minerva's flux prediction, in *Proceedings of the 10th International Workshop on Neutrino-Nucleus Interactions in Few-GeV Region (NuInt15)*. https://journals.jps.jp/doi/pdf/10.7566/JPSCP.12.010006
4. NA61/SHINE Collaboration, N. Abgrall, et al., Measurements of cross sections and charged pion spectra in proton-carbon interactions at 31 GeV/*c*. Phys. Rev. C **84**, 034 604 (2011)
5. NA61/SHINE Collaboration, N. Abgrall, et al., Measurement of production properties of positively charged kaons in proton-carbon interactions at 31 GeV/*c*. Phys. Rev. C **85**, 035 210 (2012)
6. T2K Collaboration, K. Abe, et al., T2k neutrino flux prediction. Phys. Rev. D **87**, 012 001 (2013)
7. I.T. Jolliffe, J. Cadima, Principal component analysis: a review and recent developments. Philos. Trans. R. Soc. A Math. Phys. Eng. Sci. **374**(2065), 20 150 202 (2016). https://royalsocietypublishing.org/doi/pdf/10.1098/rsta.2015.0202
8. C. Andreopoulos, et al., The GENIE neutrino monte carlo generator. Nucl. Instrum. Meth. A **614**, 87–104 (2010). arXiv: 0905.2517 [hep-ph]
9. DUNE Collaboration, B. Abi, et al., Long-baseline neutrino oscillation physics potential of the DUNE experiment. Eur. Phys. J. C **80**(10), 978 (2020). arXiv:2006.16043 [hep-ex]

10. A. Bodek, U.K. Yang, Higher twist, w scaling, and effective lo pdfs for lepton scattering in the few gev region. J. Phys. G: Nucl. Part. Phys. **29**(8), 1899 (2003). (Jul.)
11. A. Bodek, J.L. Ritchie, Fermi-motion effects in deep-inelastic lepton scattering from nuclear targets. Phys. Rev. D **23**, 1070–1091 (1981)
12. J. Nieves, J.E. Amaro, M. Valverde, Inclusive quasielastic charged-current neutrino-nucleus reactions. Phys. Rev. C **70**, 055 503 (2004)
13. T2K Collaboration, K. Abe, et al., Search for CP violation in neutrino and antineutrino oscillations by the T2K experiment with 2.2×10^{21} protons on target. Phys. Rev. Lett. **121**, 171 802 (2018)
14. MINERvA Collaboration, P.A. Rodrigues, et al., Identification of nuclear effects in neutrino-carbon interactions at low three-momentum transfer. Phys. Rev. Lett. **116**, 071 802 (2016)
15. C. Colle,et al., Extracting the mass dependence and quantum numbers of short-range correlated pairs from $A(e, e^{'} p)$ and $A(e, e^{'} pp)$ scattering. Phys. Rev. C **92**, 024 604 (2015)
16. D. Rein, L.M. Sehgal, Neutrino-excitation of baryon resonances and single pion production. Ann. Phys. **133**(1), 79–153 (1981)
17. P. Rodrigues, C. Wilkinson, K. McFarland, Constraining the GENIE model of neutrino-induced single pion production using reanalyzed bubble chamber data. Eur. Phys. J. C **76**(8), 474 (2016). arXiv:1601.01888 [hep-ex]
18. M.A. Acero, et al., New constraints on oscillation parameters from ν_e appearance and ν_μ disappearance in the NOvA experiment. Phys. Rev. D **98**, 032 012 (2018). arXiv:1806.00096 [hep-ex]
19. M. Day, K.S. McFarland, Differences in quasielastic cross sections of muon and electron neutrinos. Phys. Rev. D **86**, 053 003 (2012)

Chapter 6
Four-Channel PRISM Oscillation Analysis

The purpose of the PRISM analysis is to extract the neutrino oscillation parameters in a manner that minimises the reliance on the neutrino interaction model through data-driven predictions of the FD event rate. Chapter 4 demonstrated how a PRISM prediction of the FD event rate is produced by linearly combining off-axis measurements at the DUNE ND. The systematic uncertainties that may alter the calculation of the PRISM prediction, and therefore impact the precision of the oscillation measurement, are described in detail in Chap. 5.

This chapter details the process of producing PRISM sensitivities and presents the results for the sensitivity of PRISM to the oscillation parameters of interest to DUNE. The impact of the systematic uncertainties on the PRISM analysis sensitivities are also examined. The results presented here make use of all four oscillation signal channels accessible by DUNE and demonstrates the ability of the PRISM analysis to perform measurements of Δm^2_{32}, $\sin^2\theta_{23}$, $\sin^2 2\theta_{13}$ and δ_{CP}. A simulated measurement of the neutrino mass ordering has not been included in this thesis. The standard DUNE oscillation analysis has demonstrated that the neutrino mass ordering will be unambiguously determined within the first few years of data-taking without considering any contribution to the measurement from off-axis ND data [1]. A choice was therefore made for this chapter and Chap. 7 to focus on precision measurements of the neutrino oscillation parameter values as the best way to demonstrate how a PRISM analysis can contribute to DUNE.

6.1 Exposure at the Near and Far Detectors

In order to obtain accurate PRISM oscillation parameter sensitivities, a mapping between the POT exposures at the near and far detector sites is required. The four detector modules planned for the the FD will be installed sequentially over a number

C. M. Hasnip, *DUNE-PRISM—A New Method to Measure Neutrino Oscillations*,
Springer Theses, https://doi.org/10.1007/978-3-032-07251-1_6

Table 6.1 Mapping from years of exposure to kt-MW-Yrs. The mapping accounts for the assumed FD staging plan for module installation and the PIP-II upgrade

Years	kt-MW-Yrs
7	288
10	576
15	1056

of years. There is also the planned PIP-II upgrade that will significantly increase the power of the neutrino beam. At the time of writing, a staging plan for the FD module installation has not been finalised, therefore the following plausible FD staging plan is assumed for this thesis:

- First three years with two FD modules, providing a 20 kt fiducial mass. Initial 1.2 MW beam power.
- After three years, add a third detector module, providing a 30 kt fiducial mass.
- After five years, add the fourth detector module, reaching the total FD fiducial mass of 40 kt.
- At year-six complete the PIP-II upgrade to the neutrino beam. The beam power increases to 2.4 MW.

The exposure is expressed in units of kt-MW-Yrs, which is agnostic with regard to the particular FD staging. This unit of exposure is simply the multiplication of the fiducial mass and beam power in each year summed to the total number of years of data-taking. It is assumed that there is no lead time in increasing the power of the neutrino beam to full capacity. The mapping from the number of years of exposure to kt-MW-Yrs is shown in Table 6.1.

In addition to the assumed FD staging plan, a ND run-plan is established that maintains the ND in the on-axis position for the majority of the year followed by equal amounts of time in each off-axis position. This is the run-plan detailed in Sect. 4.2, where it is assumed that there will be 28 weeks of beam exposure each year. As described in Sect. 4.2, all the ND MC is weighted according to the ND run-plan shown in Fig. 4.8b.

6.2 NuFIT 4.0 Oscillation Parameters

To simulate a PRISM oscillation measurement for DUNE a set of true oscillation parameters is first chosen. Unless otherwise stated, the NuFIT 4.0 global fit results are assumed to be the true oscillation parameters for all the results presented in this chapter. The NuFIT 4.0 parameter values and the corresponding uncertainties are summarised in Table 6.2. The NuFIT 4.0 parameter uncertainties tend to be non-Gaussian. Therefore, for the parameters θ_{12}, $\sin^2 2\theta_{13}$ and Δm^2_{21} a Gaussian error is

Table 6.2 Best fit values and uncertainties of the oscillation parameters. Normal mass ordering is assumed. Certain parameters have their 1σ error approximated to a Gaussian uncertainty. The final column indicates whether constraints derived from the parameter uncertainty are applied to the parameter in the oscillation fit. The oscillation parameters and their uncertainties are from the 2018 NuFIT 4.0 results [2] and the Earth's density is taken from Ref. [3]

Parameter	Value and uncertainty	Constrained in fit?
θ_{12}	$33.82 \pm 0.78\ ^\circ$	Yes
$\sin^2 \theta_{23}$	$0.58^{+0.017}_{-0.021}$	No
$\sin^2 2\theta_{13}$	0.088 ± 0.003	Yes
δ_{CP}	$215^{+40}_{-29}\ ^\circ$	No
Δm^2_{21}	$7.39 \pm 0.203 \times 10^{-5}\mathrm{eV}^2$	Yes
Δm^2_{32}	$+2.4511^{+0.028}_{-0.027} \times 10^{-3}\mathrm{eV}^2$	No
Earth's density ρ	$2.848 \pm 0.057\mathrm{g/cm}^3$	Yes

approximated by calculating 1/6 of the allowed 3σ range of values. This is done in order to conveniently constrain certain parameters in the oscillation measurement that are precisely measured by other experiments. In addition, the normal mass ordering (positive Δm^2_{32}) is assumed for all sensitivities shown in this chapter.

6.3 MINUIT Oscillation Fit

The oscillation parameters are measured by fitting the PRISM prediction to the measured FD event rate. A region of oscillation parameter space is identified and a series of equidistant fit-points in the parameter space are defined. At each fit-point the MINUIT framework is used to fit the PRISM prediction to the FD data [4]. MINUIT minimises a function that quantifies the goodness of fit between the PRISM prediction and FD data and depends on a number of parameters. The minimisation is performed using the variable-metric method implemented in the MIGRAD algorithm [5]. The strategy of the MIGRAD algorithm is to iteratively vary the fit parameters by small steps until a set of parameters is found that minimises the goodness of fit function. The goodness of fit function is typically a chi-squared (χ^2) or likelihood function. This process is repeated for each fit-point, building a map of goodness of fit values across the oscillation parameter space of interest.

In the case of the PRISM analysis, the goodness of fit function to be minimised is a χ^2 function that depends on the oscillation and systematic uncertainty parameters. Figure 6.1 illustrates the extent of the agreement between the PRISM prediction to the FD data over a Δm^2_{32} and $\sin^2 \theta_{23}$ parameter space, where the true oscillation point is at the NuFIT 4.0 oscillation parameters. At each point in the oscillation parameter space the PRISM prediction is fitted to the FD data to determine the minimum χ^2. This builds a map of χ^2 values covering the whole oscillation parameter space of interest from which sensitivity contours may be plotted.

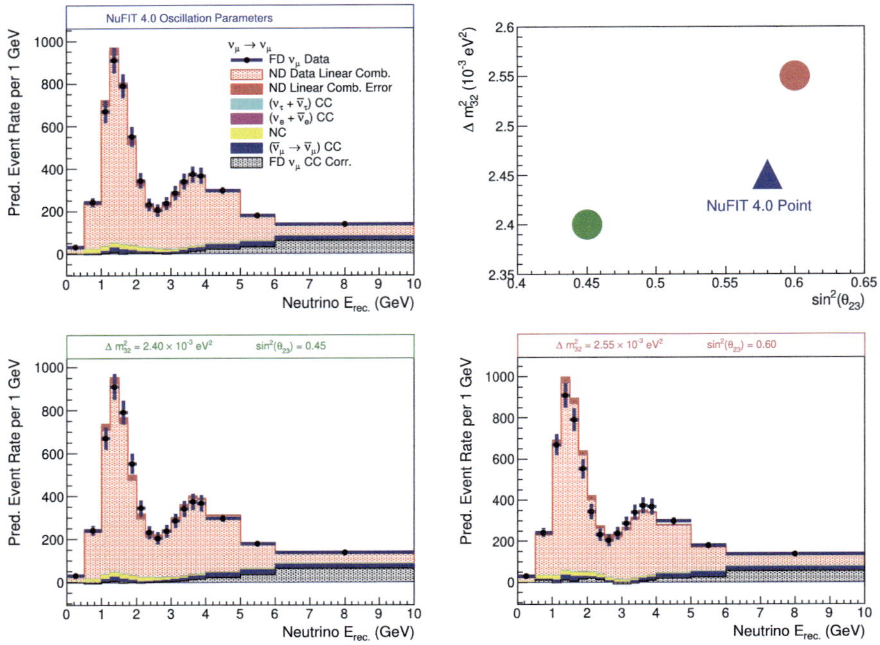

Fig. 6.1 Illustration of the agreement between the PRISM prediction to the FD data in an oscillation parameter space. The FD data in each window is set to the NuFIT 4.0 oscillation parameters. A PRISM prediction is generated at each point in the Δm^2_{32} and $\sin^2\theta_{23}$ parameter space and fitted to the FD data. At the true oscillation point (top left window) the PRISM prediction will have the best fit to the FD data, resulting in the smallest χ^2 calculation

6.4 PRISM Goodness of Fit

In order to fit the PRISM prediction to the FD data using MINUIT, a χ^2 function that depends on the oscillation and systematic parameters is defined. The chosen χ^2 function is based on the common least-squared estimator:

$$\chi^2(\theta, x) = \sum_{i}^{N_{bins}} \sum_{j}^{N_{bins}} \left(P_i(\theta, x) - D_i\right) V_{ij}^{-1} \left(P_j(\theta, x) - D_j\right), \tag{6.1}$$

where θ and x are vectors of the oscillation and nuisance parameter values included in the fit respectively. The indices i and j refer to histogram bins of the reconstructed energy variable of choice and the PRISM prediction and FD data are represented by $P_{i,j}(\theta, x)$ and $D_{i,j}$ respectively. The PRISM prediction $P_{i,j}(\theta, x)$ can be made for any set of oscillation and nuisance parameters. Finally, V_{ij}^{-1} are elements of an inverse covariance matrix. The total statistical uncertainty in the PRISM analysis is the quadrature sum of the statistical uncertainty on the Poisson-distributed FD data and the PRISM linear combination of the ND data. The elements of total covariance matrix to be used in Eq. 6.1 are therefore,

$$V_{ij} = V_{ij}^{FD} + V_{ij}^{PRISM}, \tag{6.2}$$

where V_{ij}^{FD} and V_{ij}^{PRISM} are the elements of the FD data and PRISM linear combination covariance matrices respectively. The $\mathbf{V^{FD}}$ matrix is diagonal, whilst the $\mathbf{V^{PRISM}}$ matrix is, in general, not diagonal. Examples of the PRISM linear combination covariance matrices are shown in Fig. 4.19 and details on how the covariance matrices are calculated can be found in Chap. 4.

A χ^2 test statistic is calculated instead of the Poisson-loglikelihood in order to conveniently include the statistical covariances from the PRISM linear combination. However, the traditional Pearson and Neyman calculations of χ^2 are known to introduce biases, especially at low statistics. Consequently, an approximation of the Poisson-likelihood is introduced that provides a χ^2 calculation that is less biased than the Pearson and Neyman calculations. The test statistic is called the "combined Neyman-Pearson chi-square" (χ^2_{CNP}) and details of its derivation can be found in Ref. [6]. Using the combined Neyman-Pearson chi-square, the elements of the FD covariance matrix are

$$V_{ij}^{FD} = 3/\left(\frac{1}{D_i} + \frac{2}{P_i(\theta, x)}\right)\delta_{ij}, \tag{6.3}$$

where δ_{ij} is the Kronecker delta function. In summary, the total statistical uncertainty in a PRISM analysis is determined by the sum of statistical covariances on the FD data and linearly combined ND data. This statistical uncertainty is quantified by a covariance matrix (Eq. 6.2) that is used in the calculation of a χ^2 test statistic (Eq. 6.1) that defines the goodness of fit of the PRISM prediction to the FD data.

Examining different neutrino oscillation signal channels provides sensitivity to the full range of oscillation parameters accessible to DUNE. In practice, a joint fit between all four of the signal channels is performed that maximises the sensitivity of DUNE. The joint fit requires summing the individual calculations of χ^2 for the different signal channels to give a joint χ^2 of

$$\chi^2_{Joint}(\theta, x) = \chi^2_{\nu_\mu}(\theta, x) + \chi^2_{\nu_e}(\theta, x) + \chi^2_{\bar{\nu}_\mu}(\theta, x) + \chi^2_{\bar{\nu}_e}(\theta, x), \tag{6.4}$$

where each of the contributions to $\chi^2_{Joint}(\theta, x)$ are calculated from Eq. 6.1.

Neutrino oscillation probabilities for DUNE are dominated by the parameters Δm^2_{32}, $\sin^2\theta_{23}$, $\sin^2 2\theta_{13}$ and δ_{CP}. However, the subdominant parameters, such as the solar mixing parameters, are still included in θ for the χ^2 calculation. Since these subdominant parameters are well-constrained by other neutrino oscillation experiments, additional penalty terms are included in the χ^2 function that are proportional to the parameter uncertainties defined in Table 6.2. Penalty terms are also added to constrain the variation of each systematic parameter included in x. The total χ^2 that MINUIT minimises is therefore,

$$\chi^2_{Total}(\theta, x) = \chi^2_{Joint}(\theta, x) + \sum_k^{N_{osc}} \left(\frac{\Delta\theta_k}{\sigma_k}\right)^2 + \sum_n^{N_{syst}} \left(\frac{\Delta x_n}{\sigma_n}\right)^2, \tag{6.5}$$

where $\chi^2_{Joint}(\theta, x)$ is calculated from Eq. 6.4. The second term in Eq. 6.5 is the penalty term for N_{osc} oscillation parameters, where $\Delta\theta_k$ is the change relative to the central value of kth oscillation parameter and σ_k is the prior uncertainty on that parameter, both of which are taken from Table 6.2. The parameters θ_{12}, Δm^2_{21} and the matter density of the Earth ρ are always accompanied by a penalty term in the oscillation fit. Reactor neutrino experiments provide a tight constraint on θ_{13}. Therefore, unless stated otherwise, a penalty term based on the reactor neutrino measurements is also included on θ_{13}. Parameters Δm^2_{32}, $\sin^2\theta_{23}$ and δ_{CP} are allowed to vary freely in all oscillation fits. The final term in Eq. 6.5 is a penalty term for N_{syst} nuisance parameters, where Δx_n is the change in the nth nuisance parameter and σ_n is the prior uncertainty on that parameter. The nuisance parameters are systematic uncertainty parameters.

6.5 PRISM Sensitivities

This section of the thesis will present PRISM sensitivities, where the PRISM prediction and FD "data" are generated by the nominal MC simulation. This is similar to an Asimov sensitivity produced by other neutrino oscillation analyses [7]. However, the PRISM sensitivities are referred to as "Asimov-like" because a statistically independent sample of linearly combined ND MC is being fitted to the FD MC, rather than fitting the FD MC to itself. One consequence of this is that the minimum χ^2 is, in general, not zero at the the true value of the oscillation parameters of interest. Therefore, the quantity

$$\Delta\chi^2(\theta, x) = \chi^2(\theta, x) - \chi^2_{min} \tag{6.6}$$

is calculated at each fit-point instead, where $\chi^2(\theta, x)$ is the chi-squared calculated at a particular point in the parameter space and χ^2_{min} is the smallest chi-squared calculated by MINUIT across the whole parameter space of interest. From the calculation of $\Delta\chi^2$ at each point in the oscillation parameter space, contours can be plotted. For example, values of $\Delta\chi^2 = 1$ correspond to the 1σ contour for a single parameter measurement [8].

The first part of this section will examine the effect of measuring different neutrino oscillation signal channels on the sensitivity to different parameters. This will be followed by full oscillation fits excluding systematic uncertainties, before the presentation of PRISM sensitivities including the full suite of systematic uncertainty parameters outlined in Chap. 5.

6.5.1 Chi-Square Parameter Scans

A useful tool to provide insight into how the different oscillation signal channels contribute to the measurement of Δm^2_{32}, $\sin^2\theta_{23}$, $\sin^2 2\theta_{13}$ and δ_{CP} is a so-called "χ^2-scan". This method involves calculating the χ^2 for each signal channel whilst varying one particular oscillation parameter. All oscillation parameters are set to the NuFIT 4.0 oscillation parameters and systematic parameters are at their central values. No systematic or oscillation parameters are included as parameters in the χ^2 minimisation. This means the χ^2 is only calculated from Eq. 6.1 and the vectors of oscillation and nuisance parameters (θ and x) are empty. Consequently, the χ^2-scan should not interpreted as an oscillation sensitivity as it excludes the effect of systematic uncertainties and correlations between the different oscillation parameters.

Despite these limitations, the χ^2-scan indicates which signal channels provide DUNE with sensitivity to particular oscillation parameters. This can be seen in Fig. 6.2, which shows the χ^2-scans for Δm^2_{32}, $\sin^2\theta_{23}$, $\sin^2 2\theta_{13}$ and δ_{CP}. It can be seen that the $\nu_\mu \rightarrow \nu_\mu$ and $\bar{\nu}_\mu \rightarrow \bar{\nu}_\mu$ channels drive the sensitivity to Δm^2_{32} and $\sin^2\theta_{23}$. However, sensitivity to the correct θ_{23} octant is only achieved by including the measurement of $\nu_\mu \rightarrow \nu_e$ and $\bar{\nu}_\mu \rightarrow \bar{\nu}_e$. Sensitivity to $\sin^2 2\theta_{13}$ and δ_{CP} is dominated by the $\nu_\mu \rightarrow \nu_e$ and $\bar{\nu}_\mu \rightarrow \bar{\nu}_e$ appearance channels.

6.5.2 Statistics-Only Sensitivities

The inclusion of oscillation parameters in a fit of the PRISM prediction to the FD data provides a PRISM measurement of the oscillation parameters that accounts for statistical uncertainty and correlations between the oscillation parameters. All the oscillation parameters not directly measured in the oscillation fit are included as parameters in the χ^2 calculated using Equation 6.1. A PRISM oscillation sensitivity can be produced for any chosen set of true oscillation parameters, often referred to as the "Asimov point" or "true point". A set of statistics-only PRISM sensitivities are shown in Fig. 6.3. In each oscillation parameter space contours are plotted for two different possible oscillation hypotheses and the true point is marked. As expected, the true point in each case is located in the centre of the 1σ contour. In the case of the $\sin^2 2\theta_{13}$ and δ_{CP} measurement, the external constraint on $\sin^2 2\theta_{13}$ is removed in order to demonstrate DUNE's ability to make an independent measurement of θ_{13}.

The PRISM analysis linearly combines off-axis measurements of ν_μ and $\bar{\nu}_\mu$ events. Consequently, it is important to consider the impact of statistical uncertainties in a PRISM analysis through the linear combination covariance matrix $\mathbf{V}^{\mathbf{PRISM}}$ seen in

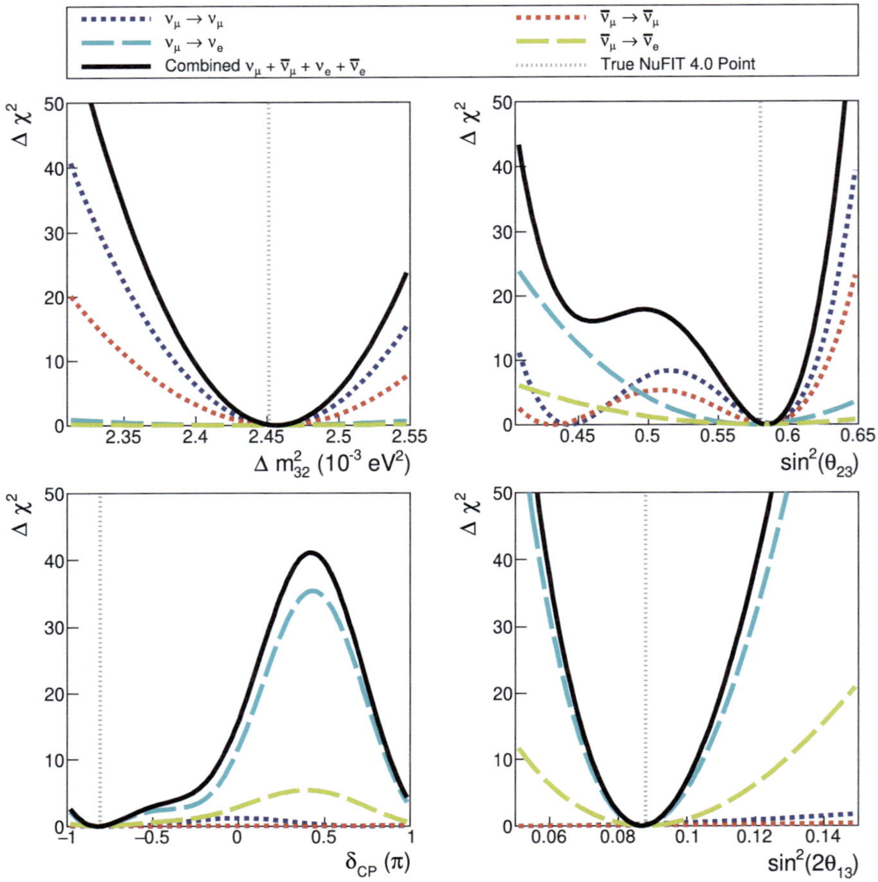

Fig. 6.2 A series of "χ^2-scans" (see text for details) for Δm^2_{32}, $\sin^2\theta_{23}$, $\sin^2 2\theta_{13}$ and δ_{CP}. The y-axis is $\Delta\chi^2(\boldsymbol{\theta}) = \chi^2(\boldsymbol{\theta}) - \chi^2_{min.}$. The contribution to the measurement of the parameters from each of the oscillation signal channels is shown. The combined χ^2, calculated by summing the χ^2 calculation from the four signal channels, is also shown

Eq. 6.2. Figure 6.4 shows the impact of statistical uncertainties at two different exposures for Δm^2_{32}, $\sin^2\theta_{23}$, $\sin^2 2\theta_{13}$ and δ_{CP}. The impact of the statistical uncertainty on the linear combination is demonstrated by including and then removing $\mathbf{V}^{\mathbf{PRISM}}$ from the χ^2 calculation. Figure 6.4 shows that the statistical uncertainties are only relevant for measurements of Δm^2_{32} and $\sin^2\theta_{23}$. However, even for these parameters and at low exposures the impact on the contour size is relatively small. As discussed in Sect. 6.5.1, measurements of Δm^2_{32} and $\sin^2\theta_{23}$ are driven by muon neutrino disappearance, whereas electron neutrino appearance dominates the sensitivity to $\sin^2 2\theta_{13}$ and δ_{CP}. The linear combination covariance matrices for each signal channel (Fig. 4.19) show that the covariances for the ν_μ and $\bar{\nu}_\mu$ predictions are, in general, one order of magnitude larger than the covariances for the ν_e and $\bar{\nu}_e$ predictions. This

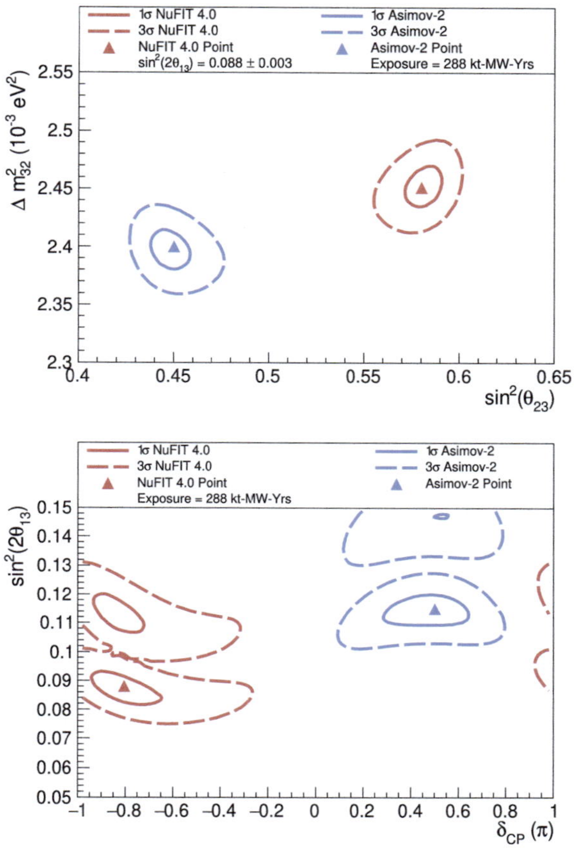

Fig. 6.3 Contours for Δm^2_{32} versus $\sin^2\theta_{23}$ (top) and $\sin^2 2\theta_{13}$ versus δ_{CP} (bottom). The sensitivities include the oscillation parameters in the fit and statistical uncertainties, however no systematic uncertainty parameters have yet been included. The penalty term on θ_{13} is removed for the $\sin^2 2\theta_{13}$ and δ_{CP} contours

is due to the muon neutrino disappearance flux being a more complex spectrum to match the ND fluxes to than the electron neutrino appearance flux, requiring larger coefficients to achieve good agreement between the linearly combined ND fluxes and the FD flux. From Fig. 6.4 it is concluded that statistical uncertainties on the linear combination of ND data, whilst important to account for in the analysis, do not significantly degrade the sensitivity of PRISM oscillation measurement.

6.5.3 Sensitivities with Systematic Uncertainties

A realistic measurement of the oscillation parameters in DUNE requires the systematic uncertainties to be accounted for in the analysis. As stated in Eq. 6.5, systematic uncertainties are implemented as nuisance parameters in the χ^2 calculation. Details on the flux, cross section and detector systematic uncertainties implemented in the PRISM analysis are presented in Chap. 5. Contours for Δm^2_{32}, $\sin^2\theta_{23}$ and

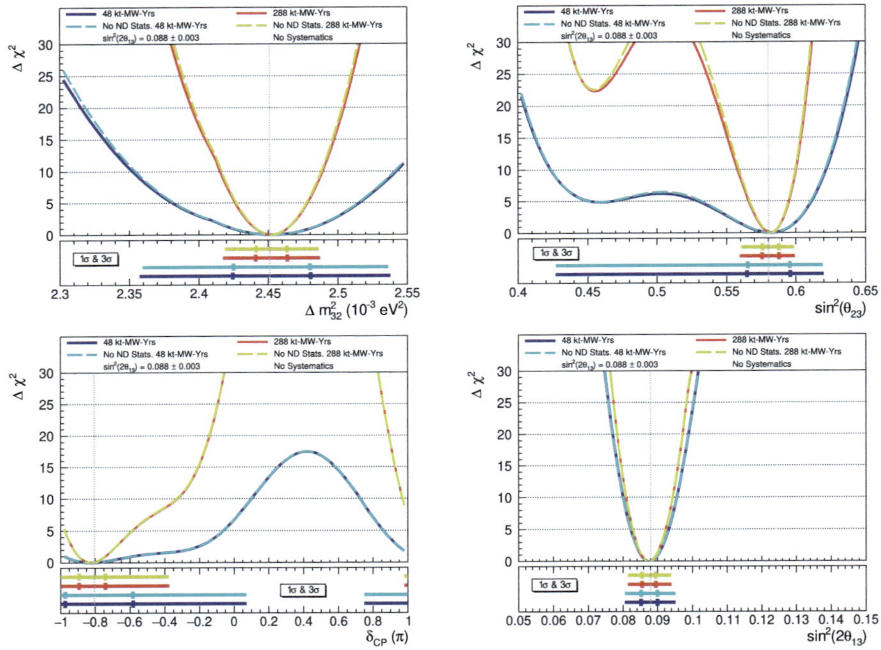

Fig. 6.4 Statistics-only fits for Δm^2_{32}, $\sin^2\theta_{23}$, $\sin^2 2\theta_{13}$ and δ_{CP} at two different exposures. The impact of the statistical uncertainty is demonstrated by removing the linear combination covariance matrix from the χ^2 calculation. The 1σ and 3σ contours are shown at the bottom of each plot. The total width of the bar is the 3σ contour and the small bars represent the 1σ contours. The true point is indicated by the vertical grey line

δ_{CP} including all systematic uncertainties are presented in Fig. 6.5. Three different exposures are shown corresponding to the FD staging plan detailed in Sect. 6.1 and Table 6.1. It can be seen in Fig. 6.5 that the width of the contours decreases as the FD exposure increases. For each oscillation fit, the true NuFIT 4.0 Asimov point, which is indicated by the grey vertical line, lies in approximately the centre of the 1σ contour. Sensitivity to the correct θ_{23} octant is achieved at 3σ at an exposure of 576 kt-MW-Yrs.

6.6 Impact of Systematic Uncertainties

A comparison between the 288 kt-MW-Yrs exposure contours in Figs. 6.4 and 6.5 demonstrates the significant impact the systematic uncertainties have in degrading the sensitivity of the PRISM oscillation analysis. It is therefore vital to understand the impact of the different flux, cross section and detector uncertainties on the oscillation parameter sensitivity. This study begins with Fig. 6.6, which shows 1σ and 3σ

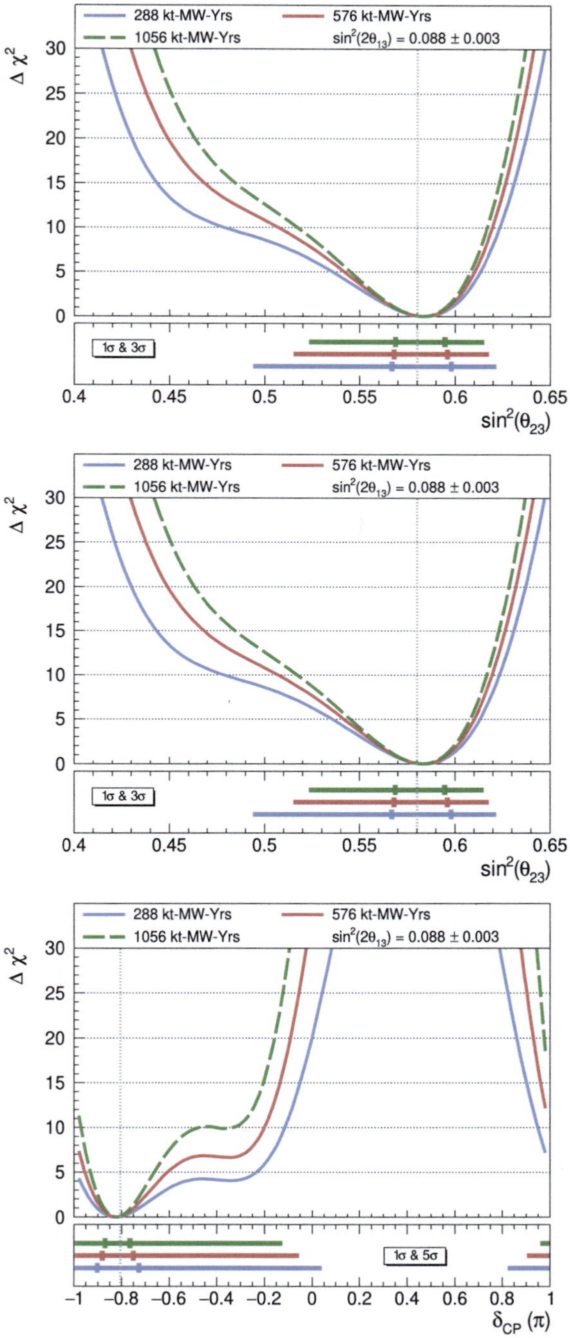

Fig. 6.5 PRISM oscillation sensitivities for three exposures and assuming the NuFIT 4.0 true oscillation parameters. All of the flux, cross section and detector uncertainties are accounted for in the fit as nuisance parameters in addition to the oscillation parameters. Penalty terms are included for θ_{12}, Δm^2_{21}, ρ and $\sin^2 2\theta_{13}$. Note that the 5σ discovery sensitivity is shown for δ_{CP}

contours for Δm^2_{32}, $\sin^2\theta_{23}$ and δ_{CP}. Contours for each parameter are made when accounting for no systematic uncertainties, only the flux systematic uncertainties, the flux and cross section systematic uncertainties and all of the systematic uncertainties. This illustrates the individual impact of the uncertainties in the flux, cross section and detector models on the sensitivity.

From Fig. 6.6 it can be seen that the flux systematic uncertainties cause a large degradation in the sensitivity to Δm^2_{32} and $\sin^2\theta_{23}$. In particular, the flux systematic uncertainties reduce the octant sensitivity by decreasing the χ^2 calculated in the region of maximal mixing around $\sin^2\theta_{23} = 0.5$. The inclusion of the cross section uncertainties also has a large impact on the contour widths of Δm^2_{32} and $\sin^2\theta_{23}$, indicating that there are components of the PRISM analysis that are dependent on the cross section model. Detector uncertainties, which are predominantly energy scale uncertainties, have a large impact on the Δm^2_{32} measurement that is comparable to the impact of the flux uncertainties. However, the degradation in the $\sin^2\theta_{23}$ sensitivity due to the detector uncertainties is less pronounced. The flux and cross section uncertainties have limited effect on the sensitivity to δ_{CP}, which is mostly affected by the detector uncertainties.

6.6.1 *Flux Systematic Uncertainty Impact*

Uncertainties in the neutrino flux model clearly have a significant impact on the PRISM sensitivity. This is not surprising; the PRISM analysis relies on the flux model in order to linearly combine the off-axis ND measurements. It was shown in Fig. 5.10 in Chap. 5 that the calculation of the linear combination coefficients is one of most important MC components through which flux uncertainties affect the PRISM analysis. This was shown to be particularly the case for the FHC predictions. The impact of the flux uncertainties through the linear combination coefficients can be demonstrated by preventing the MC predictions used to calculate the linear combination coefficients from being varied by flux systematic uncertainties in the oscillation fit. This study is shown in Fig. 6.7, where it can be clearly seen that allowing the calculation of the linear combination coefficients to be varied by flux systematic uncertainties in the fit reduces the sensitivity of the Δm^2_{32} and $\sin^2\theta_{23}$ measurements. The sensitivity to Δm^2_{32} and $\sin^2\theta_{23}$ is dominated by the measurement of $\nu_\mu \rightarrow \nu_\mu$. Therefore, this result is consistent with the findings of Fig. 5.10, which showed a large reduction in the 1σ error band after fixing the linear combination coefficient calculation in the case of the $\nu_\mu \rightarrow \nu_\mu$ prediction. In conclusion, the need to calculate a set of linear combination coefficients from the flux model in order to correctly sum many off-axis measurements at the ND is the most important means by which flux systematic uncertainties enter the PRISM analysis.

The most important flux uncertainty parameters can be determined by breaking down the flux uncertainties into the categories defined in Table 5.1. Figure 6.8 shows contours for Δm^2_{32}, $\sin^2\theta_{23}$ and δ_{CP} when including all the flux uncertainties, only the hadron production uncertainties and then the combined focusing, beam and horn

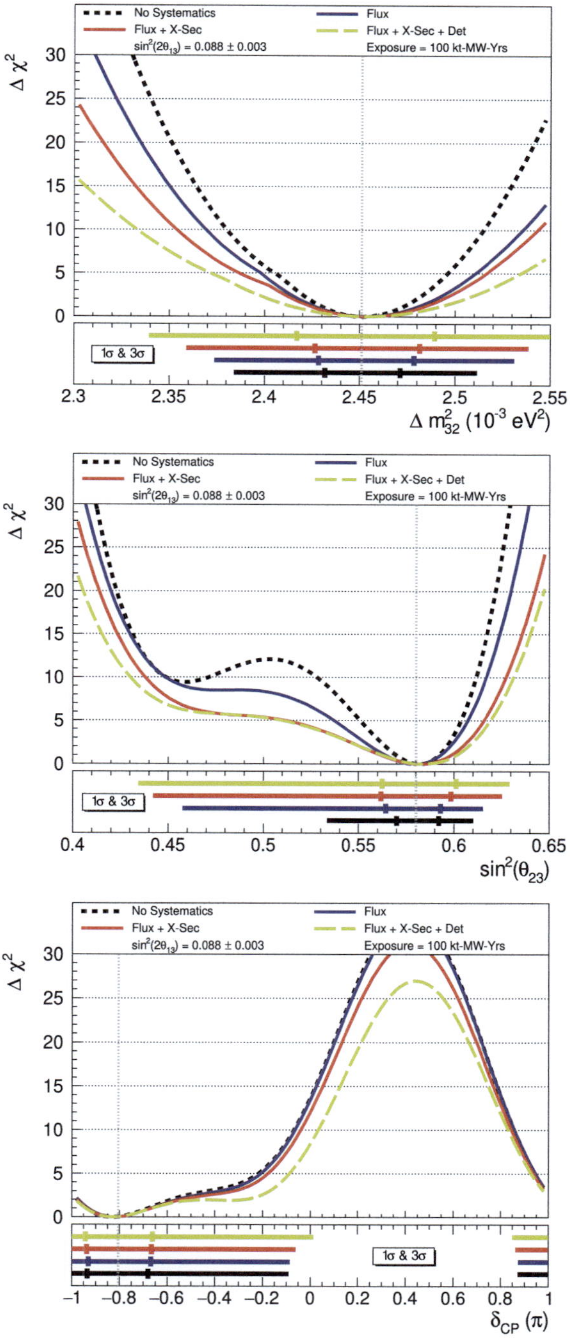

Fig. 6.6 PRISM oscillation sensitivities when accounting for no systematic uncertainty, only the flux uncertainty, the flux and cross section uncertainty and finally all the sources of systematic uncertainty. This demonstrates the relative impact of the different types of systematic parameter on Δm^2_{32} (top), $\sin^2\theta_{23}$ (middle) and δ_{CP} (bottom)

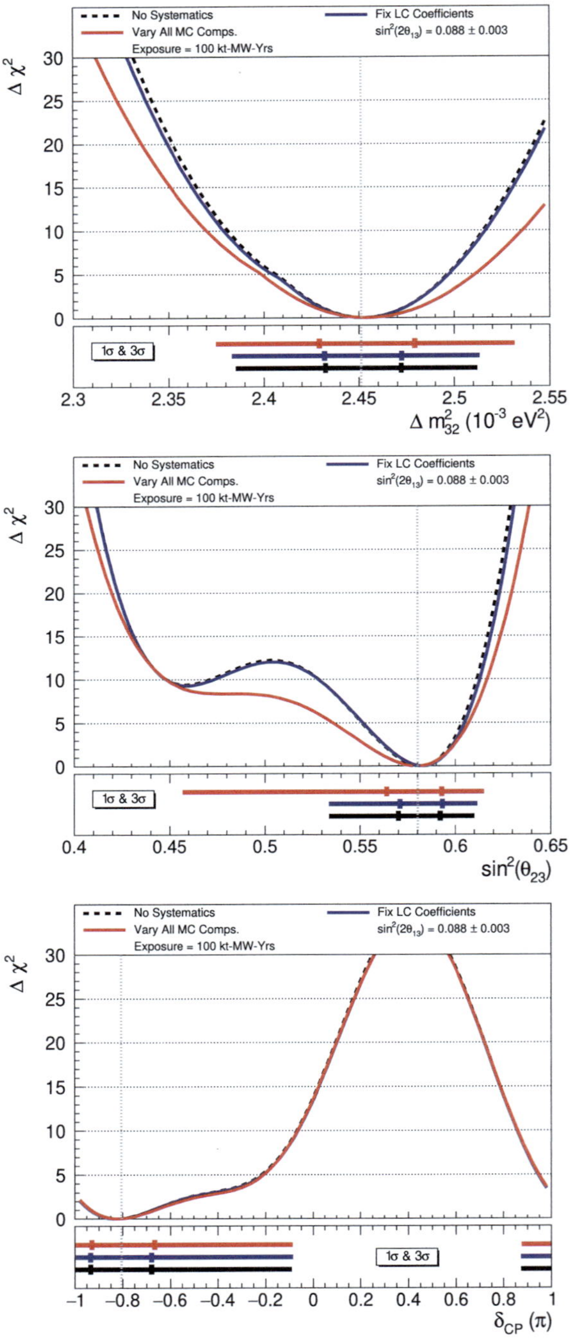

Fig. 6.7 PRISM oscillation sensitivities when accounting for no systematic uncertainty and all the flux uncertainty parameters. In the case of the blue contours, the calculation of linear combination coefficients is not allowed to vary in the fit. All the MC components of the PRISM prediction are allowed to vary for the red contours

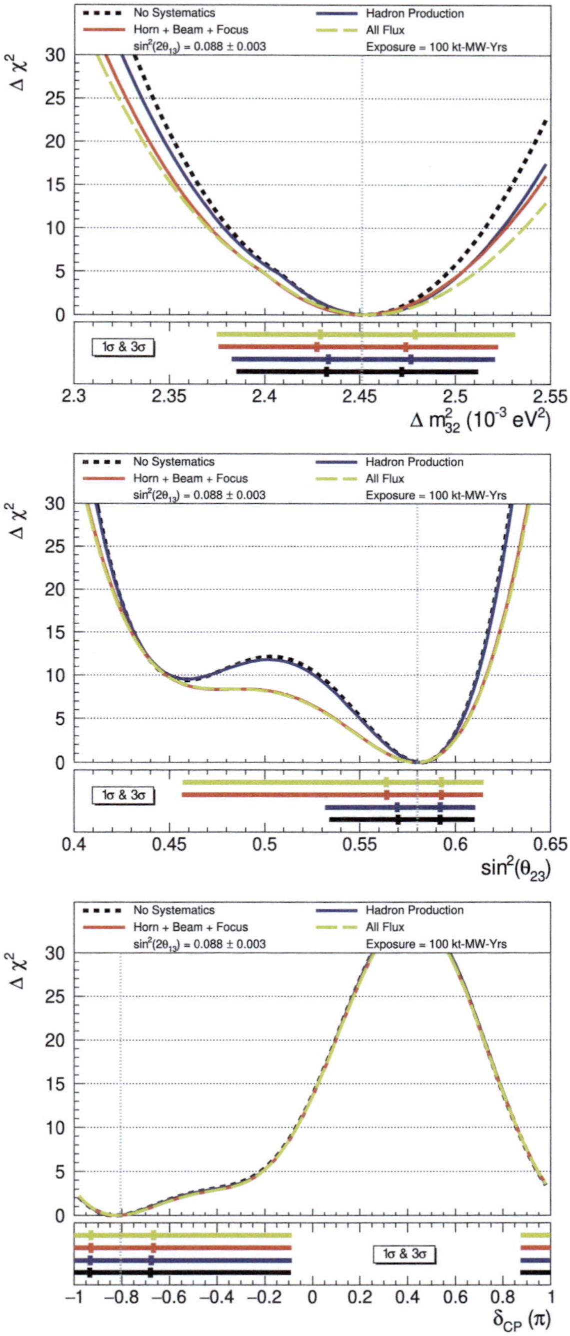

Fig. 6.8 PRISM oscillation sensitivities when accounting for no systematic uncertainty, only the hadron production uncertainty, the combined focusing and beam and horn alignment uncertainty and finally the total flux uncertainty. This demonstrates the relative impact of the different categories of flux parameter on Δm^2_{32} (top), $\sin^2\theta_{23}$ (middle) and δ_{CP} (bottom)

alignment uncertainties. It can be seen in the Δm^2_{32} contours that, whilst the hadron production uncertainties do have a significant impact on the sensitivity, the combined focusing, beam and horn alignment uncertainties are the larger effect. The hadron production uncertainties have minimal impact on the $\sin^2\theta_{23}$ contours, the width of which is entirely dominated by the focusing, beam and horn alignment uncertainties. As expected from the δ_{CP} contours shown in Fig. 6.6, none of flux uncertainty categories have a large impact on the δ_{CP} sensitivity.

It is concluded from Fig. 6.8 that, whilst the hadron production uncertainties are an important effect to consider, the focusing, beam and horn alignment uncertainties are potentially the most problematic categories of flux uncertainty to consider in a PRISM analysis. Figure 6.9 once again shows contours for Δm^2_{32}, $\sin^2\theta_{23}$ and δ_{CP}, but this time the individual impact of the focusing, beam and horn alignment uncertainties is demonstrated by only including one category of flux systematic uncertainties in the fit at a time. Horn alignment uncertainties dominate the degradation in the $\sin^2\theta_{23}$ sensitivity, particularly at $\sin^2\theta_{23} = 0.5$. Although, the focusing and beam alignment uncertainties do have a non-negligible effect in this region as well. On the other hand, the horn alignment uncertainties have little impact on the measurement of Δm^2_{32}, which is equally affected by the focusing and beam alignment uncertainties.

It has been established that the focusing, beam alignment and horn alignment flux parameters are important sources of systematic uncertainty. Each of these categories can be further investigated to identify the most problematic flux parameters. This is done through a series of so-called "N-1" studies, where one systematic parameter is removed from a set of nuisance parameters in the fit. This determines the relative contribution of a particular parameter to the overall effect of a set of nuisance parameters on the sensitivity. Four flux parameters are identified as particularly challenging for a PRISM measurement: "Horn Current", "Decay Pipe Radius", "Horn 1 X-Shift" and "Beam X-Offset".

Figure 6.10 shows the result of removing each of these parameters on a Δm^2_{32} measurement. The parameter of interest is removed from the corresponding category of flux parameters, as defined in Table 5.1. For example, in the top-right window of Fig. 6.10 the "Horn Current" parameter is removed from an oscillation fit for Δm^2_{32} that includes all of the focusing uncertainties, thereby showing the relative contribution of the "Horn Current" parameter to the combined effect of all the focusing uncertainties. The focusing parameters "Horn Current" and "Decay Pipe Radius" account for the majority of the degradation in the Δm^2_{32} sensitivity due to the focusing uncertainties. In the case of the beam alignment uncertainties, the bottom-right window of Fig. 6.10 clearly shows that "Beam X-Offset" accounts for almost all of this category's effect on the Δm^2_{32} contours.

The study for the same four flux parameters is repeated for a $\sin^2\theta_{23}$ measurement and the results are shown in Fig. 6.10. Almost all the degradation in the $\sin^2\theta_{23}$ sensitivity due to the focusing uncertainties is accounted for by "Decay Pipe Radius". Likewise, the impact of the horn alignment uncertainties is entirely accounted for by "Horn 1 X-Shift". Finally, the "Beam X-Offset" parameter completely dominates the effect of the beam alignment uncertainties on the $\sin^2\theta_{23}$ measurement.

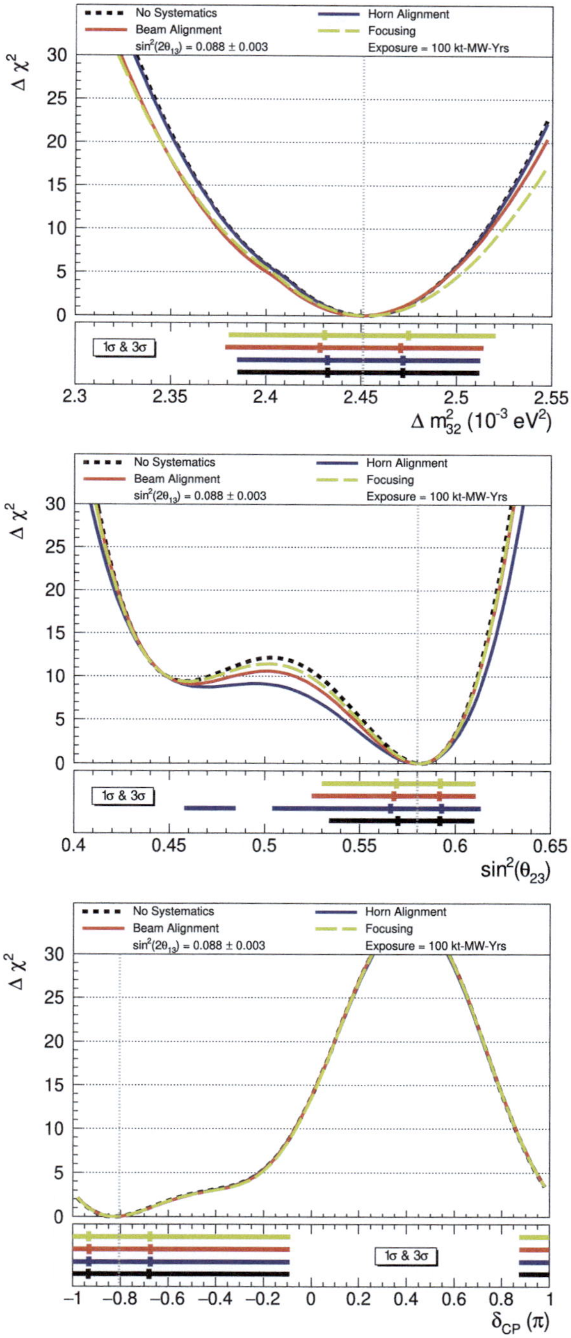

Fig. 6.9 PRISM oscillation sensitivities when accounting for no systematic uncertainty, only the focusing uncertainty, the beam alignment uncertainty and finally the horn alignment uncertainty. This demonstrates the relative impact of the different types of flux parameter on Δm^2_{32} (top), $\sin^2\theta_{23}$ (middle) and δ_{CP} (bottom)

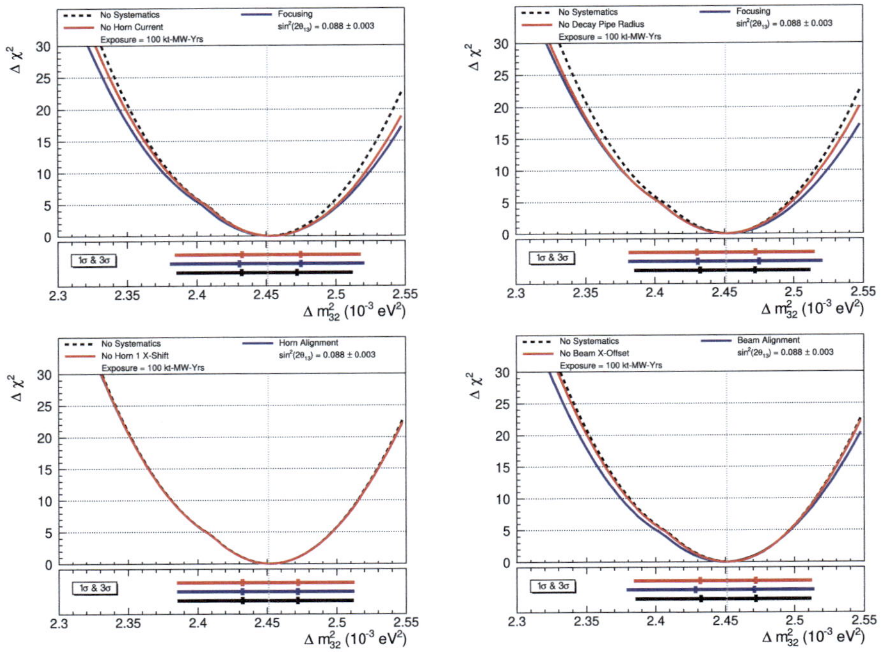

Fig. 6.10 "N-1" systematic study (see the text for details) for Δm^2_{32}. Four different flux parameters (see the legends) are removed to test the relative contribution of that parameter to the overall systematic impact of a flux uncertainty category. The 1σ and 3σ contours are shown at the bottom of each $\Delta\chi^2$ curve

Figures 6.10 and 6.11 show that the effect of the flux uncertainties on the PRISM analysis can be largely accounted for by just a handful of parameters. These parameters are: "Horn Current", "Decay Pipe Radius", "Horn 1 X-Shift" and "Beam X-Offset". It is interesting that two of the parameters—"Horn 1 X-Shift" and "Beam X-Offset"—affect the flux by altering components of the LBNF simulation in the x-direction. This is the direction in which the ND moves in a PRISM analysis. Shifts in the horn position in the y-direction have a negligible impact on the sensitivity. These two parameters were identified in Fig. 5.9 of Chap. 5 as being potentially challenging due to the large shifts in the ND off-axis flux that would not cancel with changes in the on-axis FD flux. Table 5.1 shows the uncertainties on "Horn 1 X-Shift" and "Beam X-Offset" are 0.5 mm and 0.45 mm respectively. Investigation is needed into the prospects for reducing these uncertainties. In the case of "Horn 1 X-Shift", reducing the uncertainty on movement of an approximately 2 m long object to less than half a millimetre could be challenging. But any small reduction in the tolerances for these components of the LBNF could be very useful for the PRISM analysis.

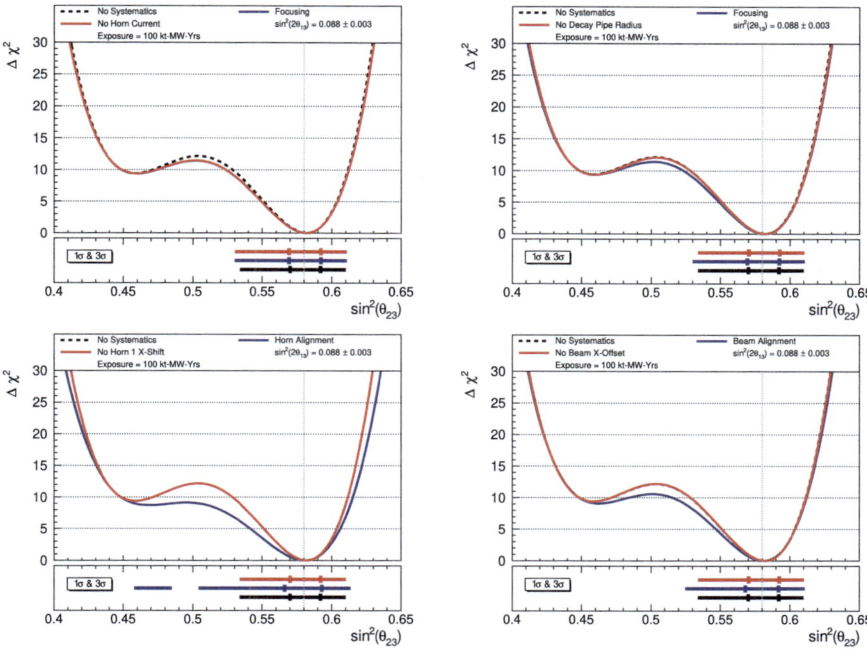

Fig. 6.11 "N-1" systematic study (see the text for details) for $\sin^2\theta_{23}$. Four different flux parameters (see the legends) are removed to test the relative contribution of that parameter to the overall systematic impact of a flux uncertainty category. The 1σ and 3σ contours are shown at the bottom of each $\Delta\chi^2$ curve

6.6.2 Cross Section Uncertainties and the Near Detector Efficiency

The PRISM analysis produces predictions of the FD event rate by linearly combining ND off-axis data. The purpose of this is to mitigate the impact of neutrino interaction modelling errors by only using the flux model to correctly combine the off-axis measurements. However, it can be seen in Fig. 6.6 that the cross section systematic uncertainties do have a significant impact on the PRISM sensitivity. Clearly, there is at least one component of the analysis that introduces a strong dependence on the neutrino interaction model to the PRISM oscillation measurement. Chapter 5 identified a likely candidate for such a component. It was shown in Fig. 5.14 that fixing the calculation of the ND efficiency correction so that it is not varied by systematic parameters results in a substantial reduction in the 1σ cross section parameter error band for the FHC predictions.

A similar study is performed for a full PRISM oscillation fit. Figure 6.12 shows the impact of accounting for the cross section uncertainties on Δm^2_{32} and $\sin^2\theta_{23}$ contours in two cases: where the ND efficiency correction is allowed to vary during the fit as normal and where the ND efficiency correction is prevented from being

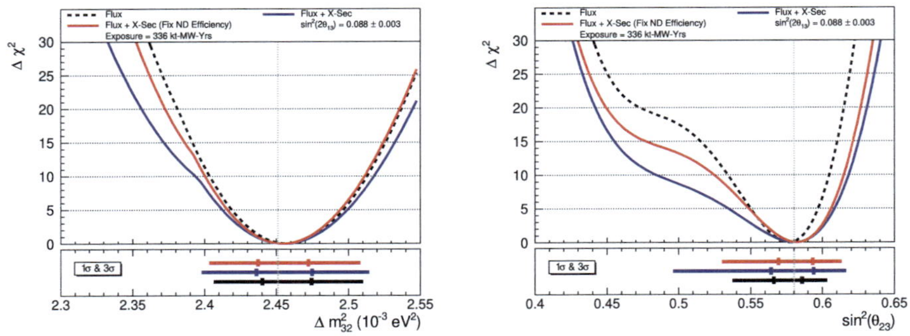

Fig. 6.12 Impact of accounting for the cross section uncertainties on the PRISM sensitivity in two cases. The blue line is the nominal case where all the MC components are affected by systemic uncertainties. The red line is the case where the ND efficiency correction calculation is not affected by systematic uncertainties. Compare the red and blue contours to the black dotted line, where only the flux uncertainties are accounted for

affected by systematic uncertainties. The difference between the red and black dotted contours is substantially less than the difference between the blue and black dotted contours. This demonstrates that a ND efficiency correction that is not affected by the cross section systematic uncertainties can bring about a significant improvement in the PRISM sensitivity.

Figure 6.12 demonstrates that removing the impact of the ND efficiency correction dramatically reduces the effect of the cross section uncertainties on the PRISM sensitivity. Therefore, it is concluded that the ND efficiency correction is the primary way cross section uncertainties enter the PRISM analysis. This is not surprising; the MC predictions of the backgrounds are relatively small (in FHC mode, at least) and the ND selection efficiency, which is based largely on acceptance cuts, is as low as 20% in some regions of phase space. Correcting for such low ND efficiencies using the MC introduces a large dependence on the neutrino interaction model. Fixing the ND efficiency calculation during the fit provides PRISM sensitivities for a scenario in which the ND efficiency is corrected in a data-driven way that does not rely on accurately modelling neutrino interactions. Details of the planned data-driven ND efficiency correction can be found in Sect. 4.7.1.

6.6.3 *Detector Systematic Uncertainty Impact*

Figure 6.6 demonstrates that detector systematic uncertainties have a significant impact on the PRISM oscillation parameter sensitivity. Once the impact of the cross section uncertainties are largely removed through the inclusion of a data-driven ND efficiency correction, the detector uncertainties are one of largest sources of systematic uncertainty in the PRISM analysis. The detector uncertainties have an impact on the Δm^2_{32} measurement of comparable size to the flux uncertainties and are the dominant source of systematic error in the δ_{CP} sensitivity. It is therefore necessary to examine which detector uncertainties cause the greatest degradation in the PRISM sensitivity to each of the oscillation parameters. This is done by accounting for one category of detector uncertainties at a time in a fit and comparing the resulting contours to the case where no systematic uncertainties are accounted for in the fit. For example, when investigating the impact of the muon energy scale uncertainty on the oscillation analysis, the p_0, p_1 and p_2 parameters (see Sect. 5.3 for details) are all included in a fit for the oscillation parameter of interest.

The impact of the total ND and FD uncorrelated energy scale uncertainties on Δm^2_{32}, $\sin^2\theta_{23}$ and δ_{CP} are shown in Fig. 6.13. It is demonstrated that the overall energy scale uncertainties for the two detectors have very little impact on the $\sin^2\theta_{23}$ and δ_{CP} contours. The ND total energy scale uncertainty also has a limited impact on the Δm^2_{32} sensitivity. However, the FD total energy scale uncertainty does have a noticeably larger effect than the ND total energy scale on the Δm^2_{32} sensitivity. This is consistent with the 1σ error band for the total energy scale shown in Fig. 5.16, where the FD energy scale error band is larger than the ND energy scale error band.

Figure 6.14 demonstrates the effect the individual particle energy scale uncertainties have on the oscillation parameter sensitivities. The most noticeable result is the large degradation in the Δm^2_{32} sensitivity due to the muon energy scale uncertainty. This is to be expected, since Δm^2_{32} sensitivity is driven by the position of the first oscillation maximum in the muon neutrino disappearance spectrum. The charged hadron and neutron energy scale uncertainties also have a large impact on the Δm^2_{32} contours. However, the EM shower energy scale uncertainty has a negligible effect on the Δm^2_{32} measurement. This is because the electron neutrino appearance events provide little additional sensitivity to Δm^2_{32}. θ_{23} determines the strength of neutrino mixing, but does not alter the position of the oscillation maximum. Hence, none of the individual particle energy scale uncertainties significantly alter the $\sin^2\theta_{23}$ contours. The EM shower energy scale uncertainty does slightly degrade the $\sin^2\theta_{23}$ octant sensitivity, since this is driven by the electron neutrino appearance measurement. Finally, the δ_{CP} sensitivity is dominated by the measurement of electron neutrino appearance events. Hence, the EM shower and charged hadron energy scale uncertainties cause the largest degradation in the δ_{CP} sensitivity.

The effect of the detector resolution uncertainties on the PRISM contours is shown in Fig. 6.15. The resolution uncertainties for the individual particle types are grouped together to demonstrate the combined impact of all the ND resolution uncertainties and all the FD resolution uncertainties on the contours. Figure 6.15 demonstrates that both the ND and FD resolution uncertainties only have a limited impact on the measurement of Δm^2_{32}, $\sin^2\theta_{23}$ and δ_{CP}. The FD resolution uncertainties have a larger impact on Δm^2_{32} and $\sin^2\theta_{23}$ than the ND resolution uncertainties. This is in agreement with Fig. 5.18, which shows the 1σ error band for FD resolution uncertainties to be larger than the error band for the ND resolution uncertainties.

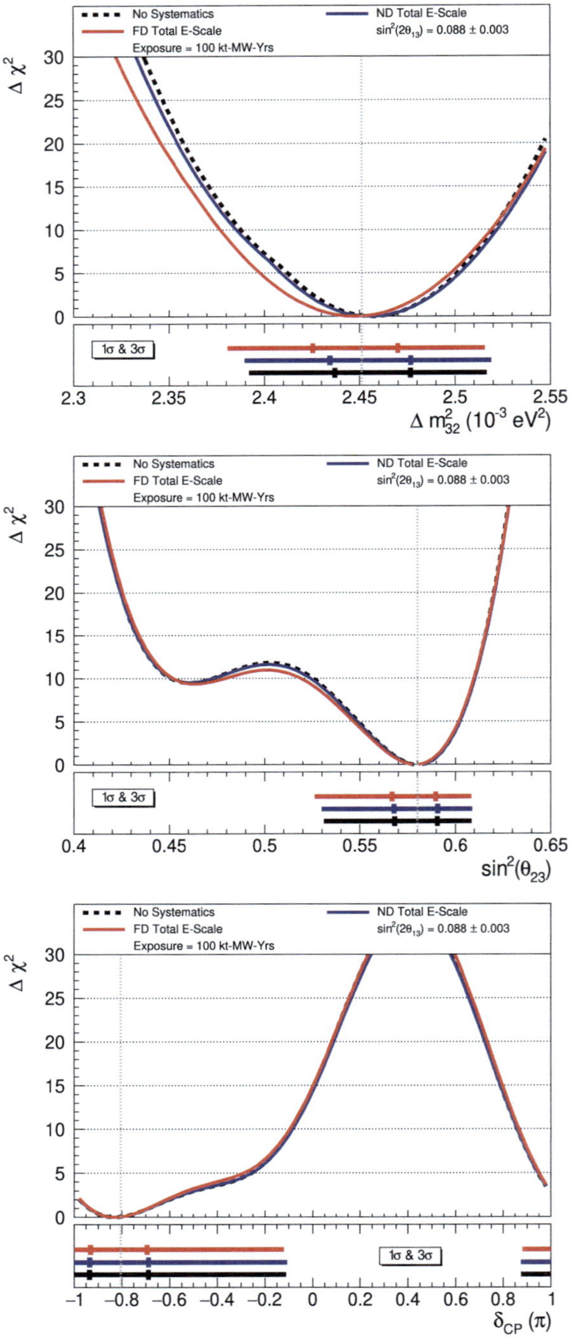

Fig. 6.13 PRISM oscillation sensitivities for Δm^2_{32}, $\sin^2\theta_{23}$ and δ_{CP} when accounting for the total energy scale uncertainties in the ND and FD. Each fit shows the combined effect of the p_0, p_1 and p_2 energy scale parameters

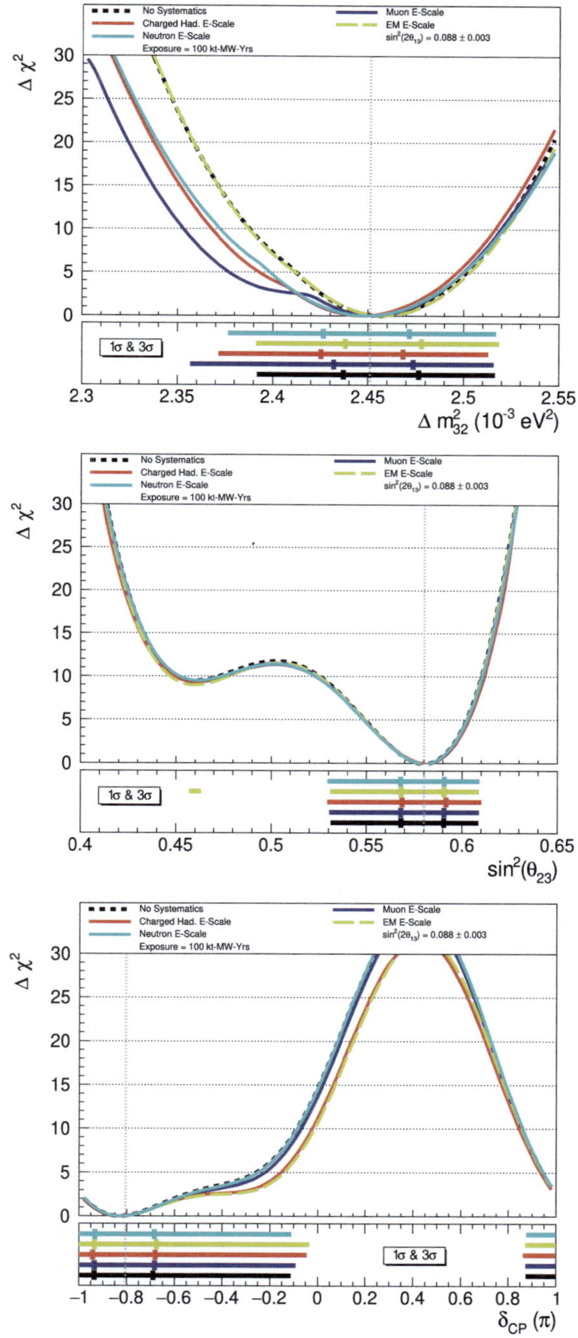

Fig. 6.14 PRISM oscillation sensitivities for Δm^2_{32}, $\sin^2\theta_{23}$ and δ_{CP} when accounting for the individual particle energy scale uncertainties. This includes the energy scale uncertainties for muon, charged hadrons, EM showers and neutrons. Each fit shows the combined effect of the p_0, p_1 and p_2 energy scale parameters

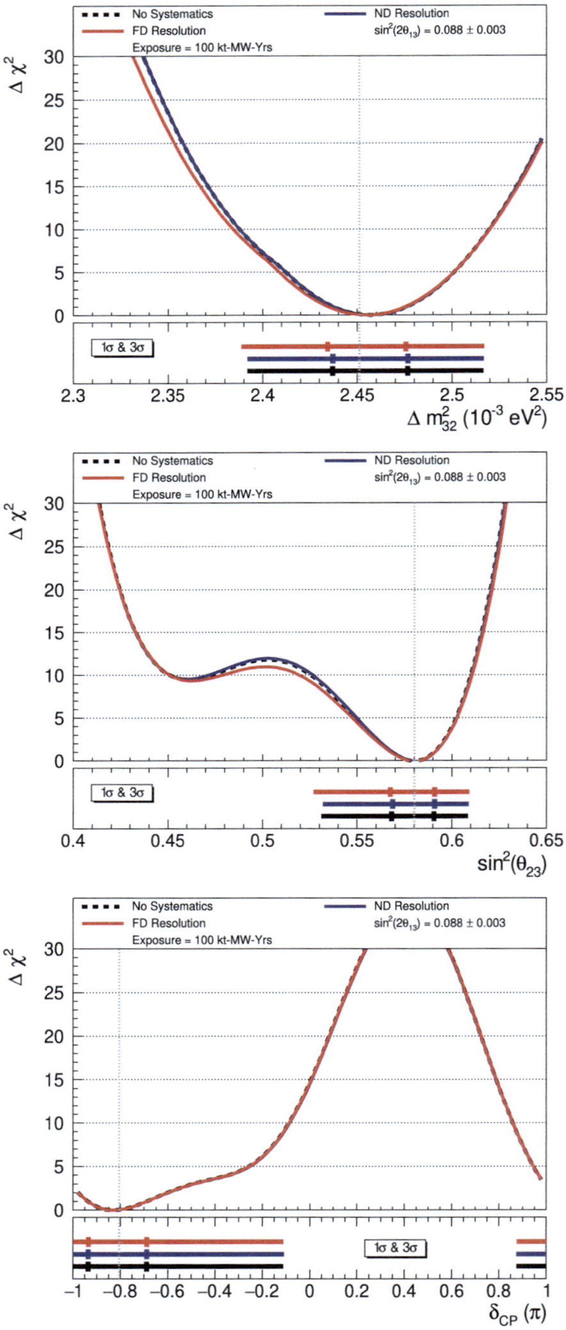

Fig. 6.15 PRISM oscillation sensitivities for Δm^2_{32}, $\sin^2\theta_{23}$ and δ_{CP} when accounting for the resolution uncertainty in the ND and FD. The coloured lines show the combined effect of all the resolution uncertainties in the ND (blue) and FD (red)

6.7 PRISM Sensitivities with a Data-Driven Near Detector Efficiency Calculation

PRISM sensitivities including all the systematic uncertainties and the four signal channels are reproduced assuming the inclusion of a data-driven ND efficiency correction. A data-driven ND efficiency correction would be calculated on an event-by-event basis directly from the ND data. The inclusion of a data-driven correction of the ND selection efficiency is planned for a future version of the PRISM analysis and more details on how this would be implemented are provided in Sect. 4.7.1. For now, this scenario is simulated by preventing the MC predictions used to calculate the ND selection efficiency from being varied by the systematic parameters.

Figure 6.16 shows the PRISM sensitivity to Δm^2_{32}, $\sin^2\theta_{23}$ and δ_{CP} assuming the NuFIT 4.0 oscillation parameters and a data-driven ND efficiency correction for three different exposures. There has been a substantial improvement in the sensitivity of the PRISM measurement of Δm^2_{32} and $\sin^2\theta_{23}$ relative to Fig. 6.5. This is entirely due to the reduction in the impact of the cross section uncertainties by fixing the ND efficiency calculation in the oscillation fit. The simulated parameter measurements and their uncertainties according to Figs. 6.16 and 6.5 in the case of data-driven and MC-based ND efficiency correction respectively are summarised in Table 6.3 for a high 1056 kt-MW-Yrs FD exposure, representing the ultimate precision the PRISM analysis can achieve without further reductions in the impact of the systematic uncertainties.

PRISM sensitivities including all systematic uncertainties are also plotted in the two-dimensional parameter spaces of Δm^2_{32} versus $\sin^2\theta_{23}$, $\sin^2\theta_{23}$ versus δ_{CP} and $\sin^2 2\theta_{13}$ versus δ_{CP}. Figures 6.17 and 6.18 show 1σ and 3σ contours at an exposure of 288 kt-MW-Yrs, which corresponds to 7-years of data according to the staging plan established in Sect. 6.1. Whilst the reactor neutrino θ_{13} constraint is applied to oscillation sensitivities shown in Fig. 6.17, it is removed for one of the $\sin^2 2\theta_{13}$ versus δ_{CP} sensitivities in Fig. 6.18. DUNE is capable of making an independent measurement of θ_{13} that is competitive with reactor neutrino experiments at high exposures [7].

The results in this chapter assume the NuFIT 4.0 oscillation parameters to be the true oscillation parameters. However, it is interesting examine the PRISM sensitivities for alternative oscillation hypotheses. For example, if the true value of δ_{CP} is far from either zero or $\pm\pi$ then the CP asymmetry in neutrino oscillations will be large. The PRISM sensitivity in such a scenario can be tested by setting $\delta_{CP} = -0.5$ and making contours for δ_{CP} at different exposures and including all the systematic parameters. This PRISM sensitivity is shown in Fig. 6.19, where δ_{CP} being either zero or $\pm\pi$ is excluded at the 5σ level by the 576 kt-MW-Yr exposure.

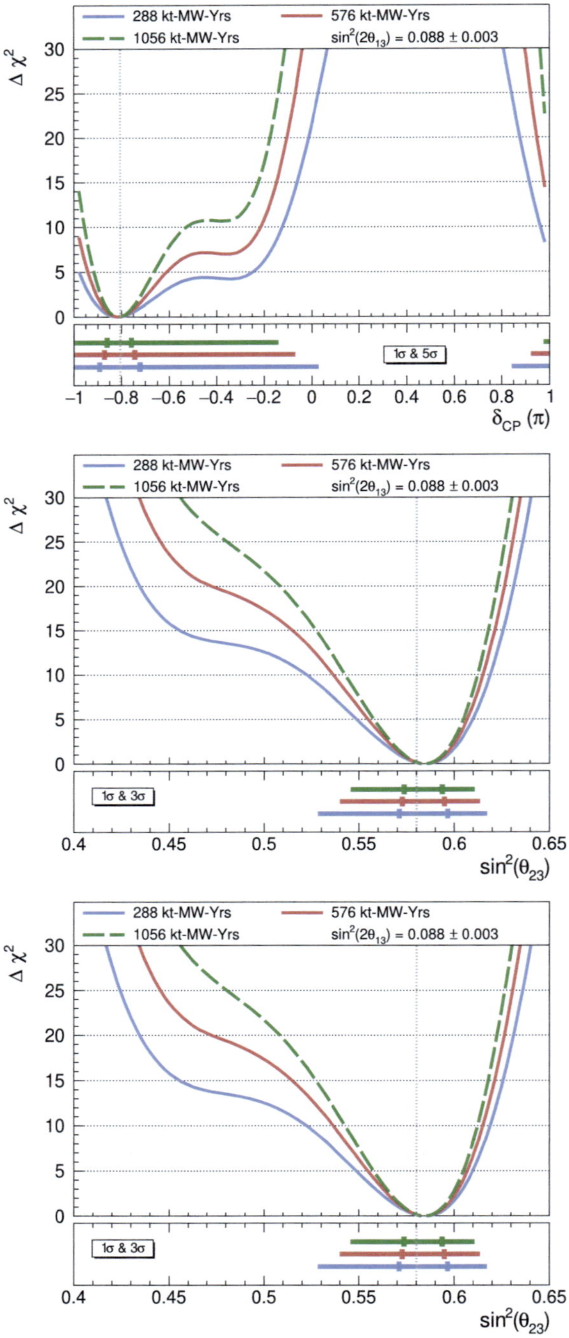

Fig. 6.16 PRISM oscillation sensitivities for three exposures and assuming the NuFIT 4.0 true oscillation parameters. All sources of systematic uncertainty (flux, cross section and detector) are accounted for. The calculation of the ND efficiency correction is fixed in all the oscillation fits. Note that the 5σ discovery sensitivity is shown for δ_{CP}

Table 6.3 Summary of the single-parameter sensitivities from Figs. 6.5 and 6.16 with the MC-based (left column) and data-driven (right column) ND efficiency correction respectively. The uncertainties refer the 1σ parameter uncertainty over a FD exposure of 1056 kt-MW-Yrs

Parameter	MC-Based ND Eff.	Data-Driven ND Eff.
Δm^2_{32}	$2.4575^{+0.0227}_{-0.0204} \times 10^{-3}$	$2.4575^{+0.0178}_{-0.0195} \times 10^{-3}$
$\sin^2\theta_{23}$	$0.5825^{+0.0121}_{-0.0139}$	$0.5825^{+0.0111}_{-0.0091}$
δ_{CP}	$-0.8200^{+0.0549}_{-0.0503}$	$-0.8200^{+0.0608}_{-0.0420}$

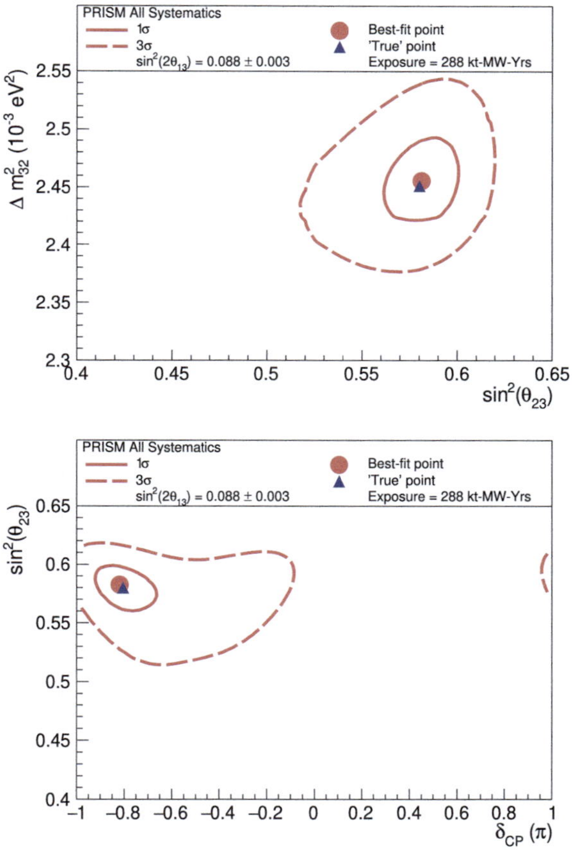

Fig. 6.17 PRISM oscillation contours at a fixed exposure in two-dimensional oscillation parameter spaces. All sources of systematic uncertainty (flux, cross section and detector) are accounted for. The reactor neutrino θ_{13} constraint is applied for both sensitivities

6.8 Comparison with On-Axis Sensitivities

Projected DUNE oscillation sensitivities using a traditional near and far detector simultaneous fit method have been published [1, 7]. This method uses on-axis ND data to constrain the flux, interaction and detector models that produce a MC prediction of the FD event rate. The methodology is colloquially referred to here as the "On-Axis" analysis. One of the primary goals of PRISM is to provide an alternative,

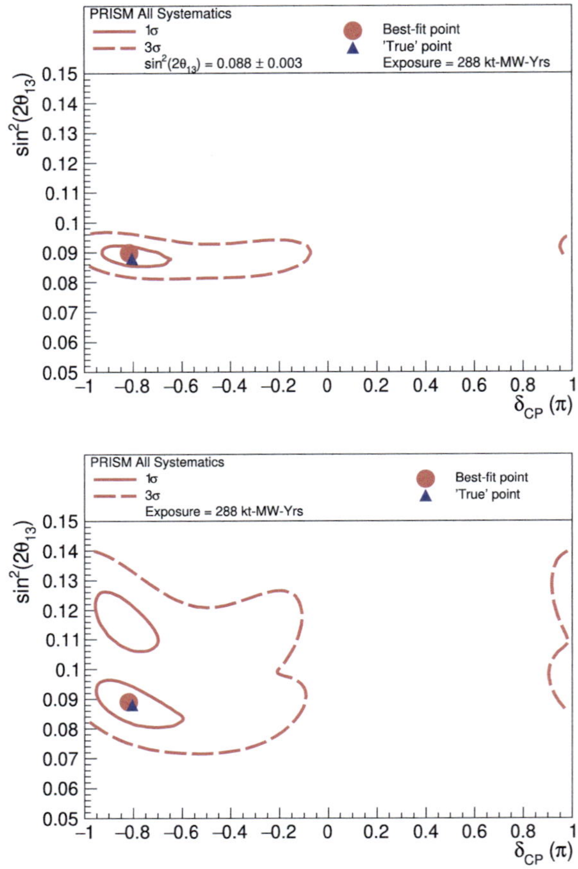

Fig. 6.18 PRISM oscillation contours for $\sin^2 2\theta_{13}$ and δ_{CP} at a fixed exposure in two-dimensional oscillation parameter spaces. All sources of systematic uncertainty (flux, cross section and detector) are accounted for. The reactor neutrino θ_{13} constraint is removed for the second (bottom) oscillation sensitivity

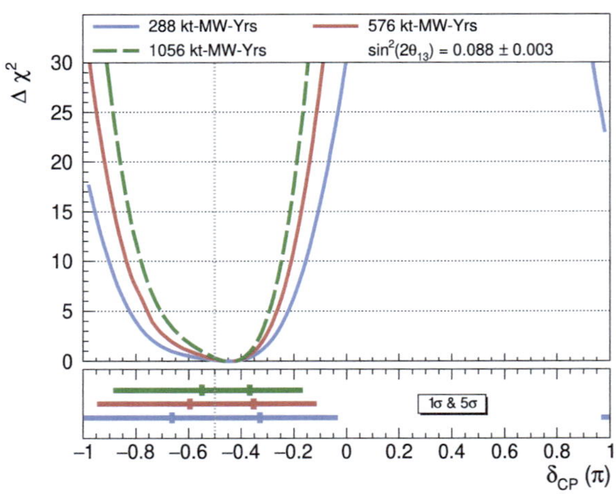

Fig. 6.19 Contours for a δ_{CP} PRISM measurement assuming a value of $\delta_{CP} = -0.5$. All systematic uncertainties are accounted for and the calculation of the ND efficiency correction is fixed

Table 6.4 Mapping from years of exposure to kt-MW-Yrs according to Refs. [1, 7]. The mapping accounts for the assumed FD staging plan and the PIP-II upgrade

Years	kt-MW-Yrs
7	336
10	624
15	1104

data-driven measurement of the oscillation parameters that is complementary to the standard On-Axis analysis. As such, it is important to compare PRISM sensitivities to the current On-Axis sensitivities and to diagnose the cause of any differences between the two analyses. However, caution should be taken in directly comparing PRISM and On-Axis sensitivities at this stage, since the On-Axis sensitivities are Asimov studies where the neutrino interaction model present in the MC perfectly describes the "data", which will not be the case in a real oscillation measurement. In this Asimov case it is expected that an on-axis ND will be effective in providing tight constraints on the cross section systematic uncertainties.

Near-equivalent exposures, systematic uncertainty models and analysis variables are needed in order to make direct comparisons between PRISM and On-Axis analysis sensitivities. To achieve this, the FD staging plan for the PRISM contours defined in Sect. 6.1 is changed to the more optimistic scenario assumed in Ref. [7]. The new mapping between the number of years of data and kt-MW-Yrs units is shown in Table 6.4. The analysis variable used in the PRISM and On-Axis analysis is the "reconstructed neutrino energy" defined in Eq. 4.1. The flux and cross section systematic models for the PRISM and On-Axis analyses are identical. There are some differences in the implementation of the flux uncertainties due to the need to accommodate off-axis fluxes in the PRISM analysis. However, the flux parameters and the underlying flux simulation are the same.

Slightly different detector uncertainty models are used in the PRISM and On-Axis analysis. The On-Axis analysis presented in Ref. [7] conservatively assumes all of the energy scale uncertainties to be wholly uncorrelated between the near and far detectors. In contrast, the PRISM analysis presented here assumes that the energy scale uncertainties for individual particle responses are correlated between the near and far detectors. The On-Axis analysis also includes parameters that quantify the uncertainty in CVN particle identification accuracy in the FD selection, although these parameters are not expected to have a significant impact on the sensitivity. In summary, the PRISM detector uncertainty model can be regarded as less conservative than the On-Axis model. However, both of these models are highly approximate and only aim to provide an initial estimate of the impact of the detector uncertainties.

A first comparison of the PRISM and On-Axis contours when accounting for all systematic uncertainties is shown in Fig. 6.20. The PRISM sensitivities shown in Fig. 6.20 have the calculation of the ND efficiency correction fixed in the fit, providing an estimate of the future PRISM contours that include a data-driven ND

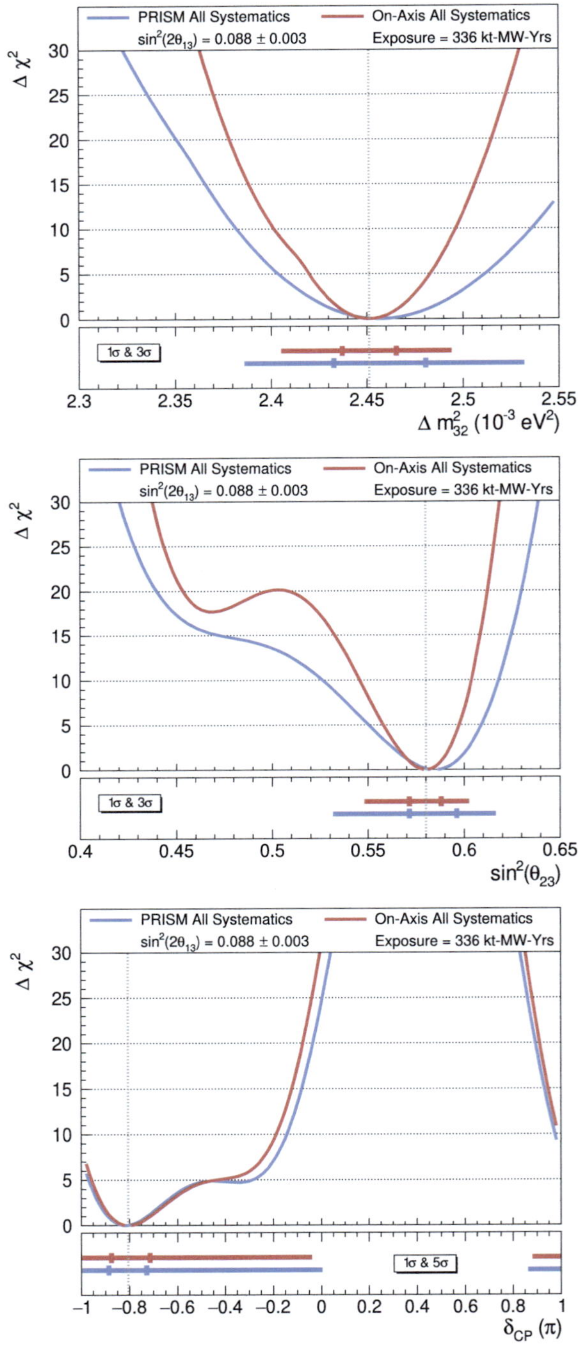

Fig. 6.20 Comparison between PRISM and On-Axis analysis contours for a fixed exposure and accounting for all sources of systematic uncertainty. The ND efficiency correction has been fixed in the PRISM oscillation fits, simulating the inclusion of a data-driven ND efficiency correction

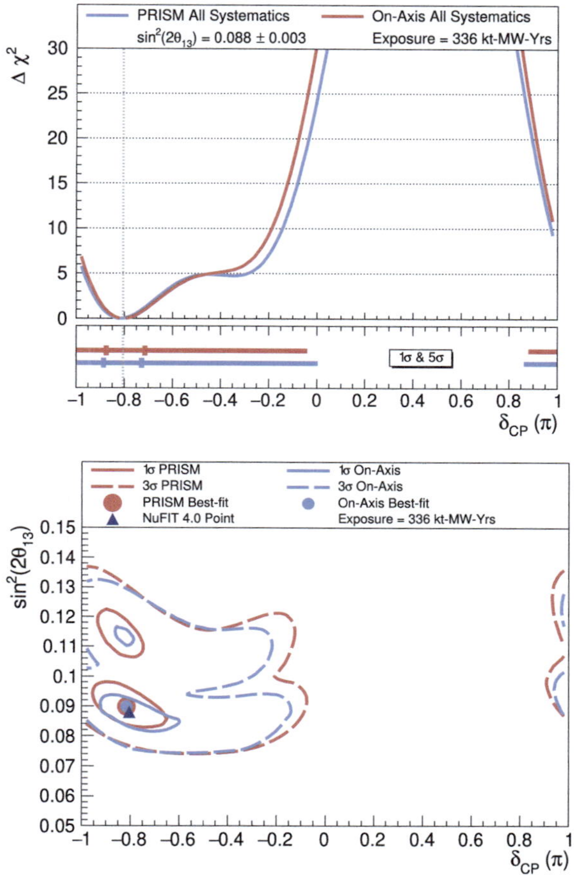

Fig. 6.21 Comparison between the PRISM (red) and On-Axis (blue) oscillation contours at a fixed exposure in two-dimensional oscillation parameter spaces. The reactor neutrino θ_{13} constraint is removed for the $\sin^2 2\theta_{13}$ versus δ_{CP} sensitivity. All sources of systematic uncertainty are accounted for in the sensitivities. The ND efficiency correction has been fixed in the PRISM oscillation fits, simulating the inclusion of a data-driven ND efficiency correction

efficiency correction. In the case of the Δm^2_{32} and $\sin^2\theta_{23}$ contours, a clear deficit can be seen in the PRISM sensitivity relative to the On-Axis sensitivity. The sensitivity to δ_{CP} is roughly equivalent for the PRISM and On-Axis contours. Figure 6.21 again shows comparisons between the PRISM and On-Axis contours, this time in two-dimensional parameter spaces for Δm^2_{32} versus $\sin^2\theta_{23}$ and $\sin^2 2\theta_{13}$ versus δ_{CP}. As would be expected from Fig. 6.20, the contours for the Δm^2_{32} versus $\sin^2\theta_{23}$ parameter space are noticeably wider than in the On-Axis analysis case. Likewise, the $\sin^2 2\theta_{13}$ versus δ_{CP} contours are very similar in the PRISM and On-Axis analyses.

The source of deficit in the PRISM sensitivity can be diagnosed by again removing certain categories of systematic uncertainty from the oscillation fits. The PRISM and On-Axis analysis contours are compared when iteratively accounting for the cross section, flux and then detector uncertainties. Figures 6.22, 6.23 and 6.24 show this study for Δm^2_{32}, $\sin^2\theta_{23}$ and δ_{CP} respectively. The ND efficiency correction is again prevented from being varied by the systematic uncertainties. It should be noted in Fig. 6.22 that the PRISM contours are wider than the On-Axis contours before

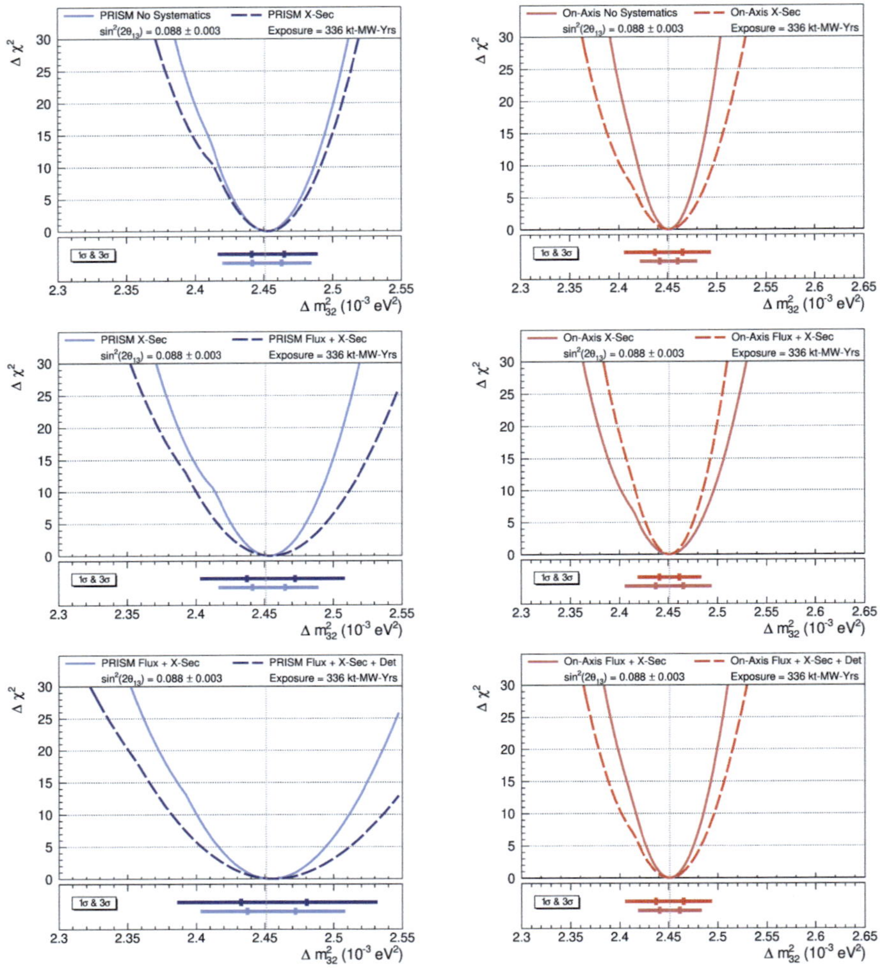

Fig. 6.22 Contours for Δm^2_{32} when accounting for the cross section (top), flux (middle) and detector (bottom) uncertainties. The impact of these systematic uncertainties on the PRISM (left) and On-Axis (right) contours are to be compared

any systematic uncertainties have been accounted for. This is due to the statistical uncertainty on the PRISM linear combination and that fact that the PRISM prediction does not perfectly match the FD "data" at the true oscillation point. This is in contrast to the On-Axis analysis, where the FD prediction is derived from the MC and therefore has negligible statistical uncertainty and exactly matches the FD "data" at the true oscillation point.

For each oscillation parameter, the cross section uncertainties have a smaller impact on the PRISM sensitivity than on the On-Axis sensitivity. However, when both the cross section and flux uncertainties are accounted for in the oscillation fit

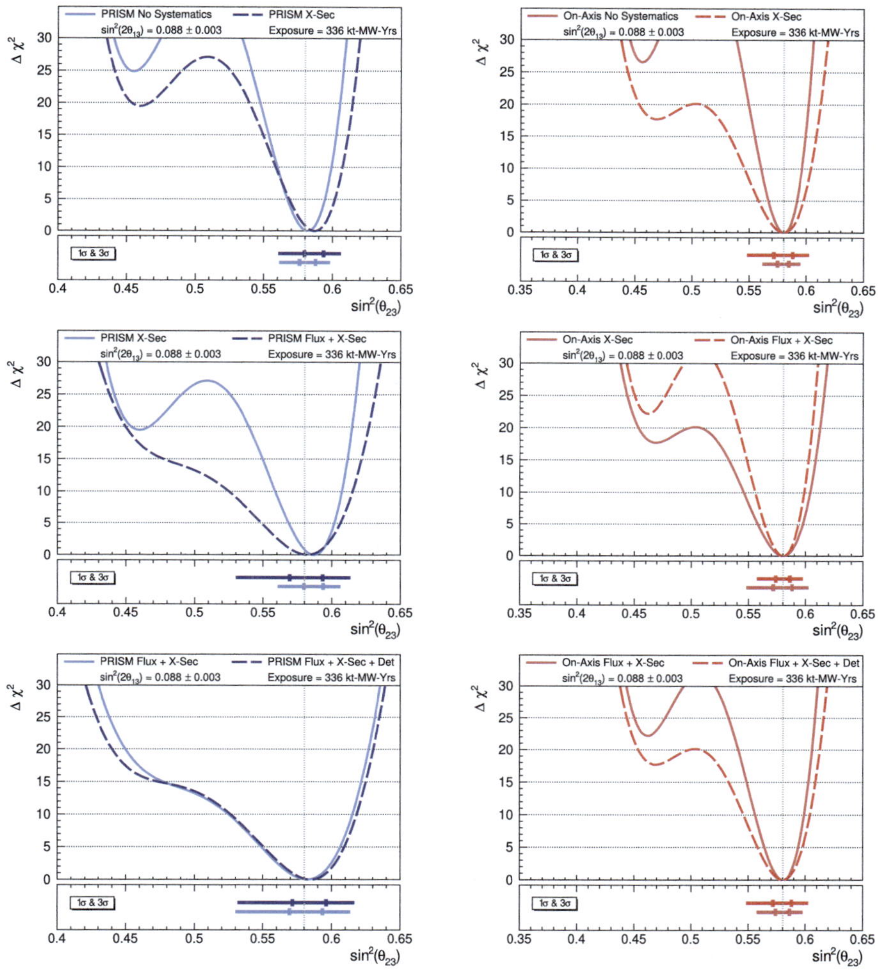

Fig. 6.23 Contours for $\sin^2\theta_{23}$ when accounting for the cross section (top), flux (middle) and detector (bottom) uncertainties. The impact of these systematic uncertainties on the PRISM (left) and On-Axis (right) contours are to be compared

the On-Axis sensitivity improves, whilst the PRISM sensitivity further degrades. This is because the On-Axis analysis measures the event rate at the ND, which is the convolution of the neutrino flux and interaction cross section. Hence, the On-Axis analysis can better-constrain the combined set of flux and cross section systematic uncertainties than either the flux or cross section uncertainties separately. On the other hand, instead of using the ND to constrain model parameters, the PRISM analysis linearly combines ND data and must accept the impact any remaining model dependence may have on the oscillation measurement. Although the impact of the cross section systematic uncertainties is small in a PRISM analysis, any flux

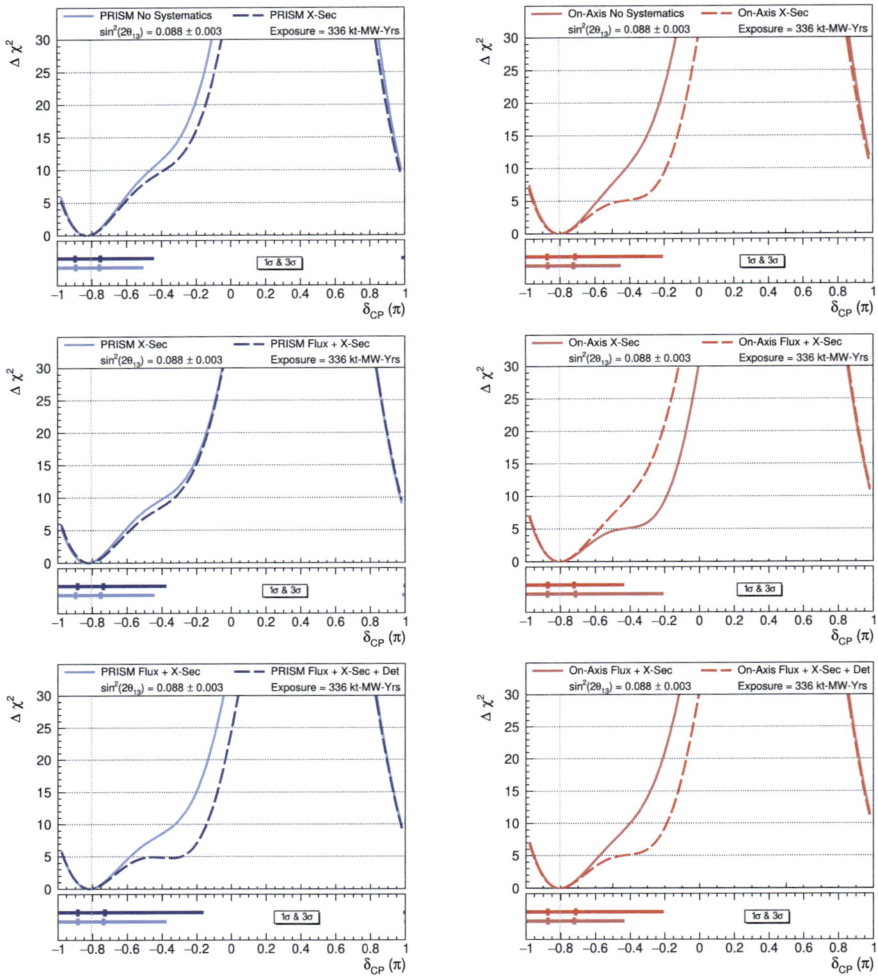

Fig. 6.24 Contours for δ_{CP} when accounting for the cross section (top), flux (middle) and detector (bottom) uncertainties. The impact of these systematic uncertainties on the PRISM (left) and On-Axis (right) contours are to be compared

uncertainties that do not cancel well between the ND and FD can have a substantial effect on the PRISM sensitivity.

The final two plots in Figs. 6.22, 6.23 and 6.24 compare the effect of detector uncertainties on the PRISM and On-Axis sensitivities for Δm^2_{32}, $\sin^2\theta_{23}$ and δ_{CP} respectively. The detector uncertainties have a similarly large impact on the PRISM and On-Axis sensitivity to Δm^2_{32} and δ_{CP}. The increase in the Δm^2_{32} PRISM contour due to the detector uncertainties is marginally larger than in the On-Axis analysis. In both analyses, the detector uncertainties have a limited effect on the $\sin^2\theta_{23}$ since the detector uncertainties are largely energy scale uncertainties.

6.9 Summary of PRISM Sensitivities

In summary, this chapter has presented a complete DUNE oscillation analysis that precisely measures the values of the oscillation parameters using the PRISM method. The rapidity with which the On-Axis analysis is expected to achieve sensitivity to the neutrino mass ordering led this thesis to focus instead on precision measurements of Δm^2_{32}, $\sin^2\theta_{23}$, $\sin^2 2\theta_{13}$ and δ_{CP}. However, a PRISM mass ordering measurement is entirely possible and should be demonstrated in a future study.

The most important sources of systematic uncertainty have been identified to be the flux and detector models. The impact of the cross section uncertainties is expected to be drastically reduced by the inclusion of a data-driven ND efficiency correction. The planned data-driven ND geometric efficiency correction is described in detail in Sect. 4.7.1. The overall effect of the flux uncertainties is dominated by just four flux parameters: "Horn Current", "Decay Pipe Radius", "Horn 1 X-Shift" and "Beam X-Offset". However, the hadron production uncertainties are still an important contribution. Alongside the flux uncertainties, the detector model is one of the most important sources of systematic uncertainty in a PRISM analysis, even when only considering detector energy scales and resolutions. As seen in Sect. 6.6.3, this is particularly the case for a Δm^2_{32} measurement, which is highly dependent on the detector energy scale. The impact the systematic uncertainties have on the precision of the oscillation parameter measurement over the lifetime of DUNE is illustrated in Fig. 6.25, which plots the width of the 1σ sensitivity band from one-dimensional PRISM oscillation fits as a function of FD exposure. The rate at which the precision improves gradually reduces over time. By 300 kt-MW-Yrs, which corresponds to just over 7-years of FD data according the staging plan described in Sect. 6.1, the rate of improvement becomes slow for both Δm^2_{32} and $\sin^2\theta_{23}$ and this is due to the impact of the systematic uncertainties on the sensitivity.

Whilst the impact of the detector uncertainties are of comparable size in the PRISM and On-Axis analyses, the flux uncertainties cause a much larger degrada-

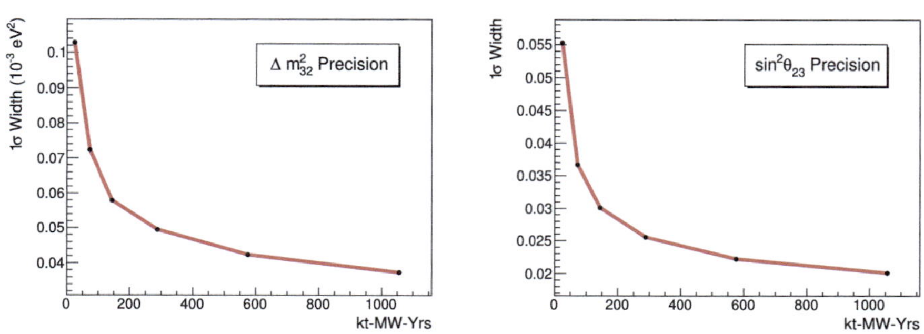

Fig. 6.25 Improvement in the precision of PRISM Δm^2_{32} and $\sin^2\theta_{23}$ measurements with increasing FD exposure. Precision is quantified by the width of the 1σ sensitivity band from a single-parameter PRISM fit. A smaller 1σ sensitivity band corresponds to a more precise measurement. The staging plan described in Sect. 6.1 is assumed

tion in the PRISM analysis sensitivities than in the On-Axis analysis sensitivities. It is concluded from Sect. 6.8 that the systematic uncertainty in the neutrino flux is the primary reason why the PRISM analysis is currently less sensitive to the oscillation parameters than the On-Axis analysis. It should be noted that the On-Axis sensitivities represent an optimistic (and unlikely) scenario where the neutrino interaction model present in the FD MC prediction accurately describes the "data". The PRISM sensitivities on the other hand are largely independent of the neutrino interaction model, whether it is an accurate description of the data or not.

This chapter has shown an Asimov oscillation analysis where the "data" and "prediction" are derived from the same MC simulations [9]. However, the purpose of the PRISM is to be able to correctly measure the oscillation parameters when the MC predictions do not accurately describe the measured neutrino event rates. A study simulating such a scenario is presented in the following chapter.

References

1. DUNE Collaboration, B. Abi, et al., Deep underground neutrino experiment (DUNE), far detector technical design report, vol. II. DUNE Physics (Feb 2020). arXiv:2002.03005 [hep-ex]
2. I. Esteban, et al., Global analysis of three-flavour neutrino oscillations: synergies and tensions in the determination of θ_{23}, δ_{CP}, and the mass ordering. JHEP **01**, 106 (2019). arXiv:1811.05487 [hep-ph]
3. B. Roe, Matter density versus distance for the neutrino beam from Fermilab to Lead, South Dakota, and comparison of oscillations with variable and constant density. Phys. Rev. D **95**(11), 113 004 (2017). arXiv:1707.02322 [hep-ex]
4. F. James, MINUIT function minimization and error analysis: reference manual version 94.1 (1994)
5. W.C. Davidon, Variable metric method for minimization. SIAM J Opti **1**(1), 1–17 (1991). https://doi.org/10.1137/0801001
6. X. Ji, et al., Combined neyman–pearson chi-square: an improved approximation to the Poisson-likelihood chi-square, in *Nuclear Instruments and Methods in Physics Research Section A: Accelerators, Spectrometers, Detectors and Associated Equipment*, vol. 961 (2020), pp. 163–677
7. DUNE Collaboration, B. Abi, et al., Long-baseline neutrino oscillation physics potential of the DUNE experiment. Eur. Phys. J. C **80**(10), 978 (2020). arXiv:2006.16043 [hep-ex]
8. Particle Data Group, P.A. Zyla, et al., Review of particle physics, in *Progress of Theoretical and Experimental Physics*, vol. 2020, no. 8 (Aug 2020), p. 083C01. https://academic.oup.com/ptep/article-pdf/2020/8/083C01/34673722/ptaa104.pdf
9. G. Cowan, K. Cranmer, E. Gross, O. Vitells, Asymptotic formulae for likelihood-based tests of new physics. Eur. Phys. J. C **71**, 1554 (2011). arXiv:1007.1727 [physics.data-an]. (Erratum: Eur.Phys.J.C 73, 2501 (2013))

Chapter 7
Oscillation Measurement Biases from the Neutrino Interaction Model

The previous chapter detailed a realistic PRISM oscillation analysis, showing the sensitivity of PRISM to Δm^2_{32}, $\sin^2\theta_{23}$, $\sin^2 2\theta_{13}$ and δ_{CP}. However, this was an "Asimov-like" study, where the near and far detector "data" were identical to the MC production. In a real DUNE analysis the data will be whatever is measured by the detectors and the MC simulations will not necessarily provide an accurate model for every aspect of the data. This is particularly the case for the neutrino interaction model and the challenges encountered in modelling neutrino interactions on nuclei are detailed in Sect. 2.6.

The default DUNE oscillation analysis (On-Axis analysis) uses a stationary on-axis ND to constrain and tune the neutrino interaction model so that the ND MC simulation matches the ND data well. There are potentially a number of degenerate ways the model parameters could be altered to achieve good agreement between the simulation and data. In addition, the model may require additional parameters to force agreement between the ND MC and data. Whilst it is preferable that any additional parameters are motivated by physical effects not yet considered in the base model, often these are empirical corrections. The predicted event rate is a convolution of the flux, interaction and detector model predictions. If the wrong choice of modifications to the neutrino interaction model are made, an inaccurate mapping from the true neutrino energy to the reconstructed energy will be extrapolated to the FD, where a new prediction of the FD event rate is made with a neutrino flux very different from the ND due to neutrino oscillations. Consequently, it is possible to have a biased prediction of the FD event rate, and therefore a biased oscillation measurement, despite achieving a satisfactory fit at the on-axis ND.

The purpose of this chapter is to provide a case study that demonstrates how a PRISM analysis can naturally avoid this scenario. PRISM builds FD event rate predictions by linearly combining off-axis ND data. Since both the ND and FD are liquid argon targets, the neutrino interaction physics should be identical in the FD data and the PRISM linearly combined prediction. In principle, this will significantly reduce

C. M. Hasnip, *DUNE-PRISM—A New Method to Measure Neutrino Oscillations*, Springer Theses, https://doi.org/10.1007/978-3-032-07251-1_7

the risk of obtaining a biased measurement of the neutrino oscillation parameters. This study builds on the neutrino oscillation measurement bias study presented in Chap. 4 of the Near Detector Conceptual Design Report [1].

7.1 Deposited Energy Analysis Variable

The analysis variable used in the previous chapters was the reconstructed neutrino energy. However, as explained in Sect. 3.6, full reconstruction is only available for the FD MC production, necessitating the use of different analysis variables in the ND and FD data. For the case study presented in this chapter a new variable is defined that is applicable to both the near and far detectors. The variable is called "deposited energy" and is given by

$$E^{dep.} = E^{true}_{lep.} + E^{dep.}_{p} + E^{dep.}_{\pi^{\pm}} + E^{dep.}_{\pi^{0}} + E^{dep.}_{other}, \tag{7.1}$$

where $E^{true}_{lep.}$ is the true lepton energy and $E^{dep.}_{p}$, $E^{dep.}_{\pi^{\pm}}$, $E^{dep.}_{\pi^{0}}$ and $E^{dep.}_{other}$ refer to the deposited energy of the protons, charged pions, neutral pions and other visible hadrons and their progeny respectively. $E^{dep.}$ therefore assumes the lepton to be perfectly reconstructed and applies no model-based corrections to the hadronic energy. It is also assumed that there are no detection thresholds for the hadrons and energy deposits are always associated with the correct final state particle. The selection cuts used in the ND and FD and the resulting background samples are identical to those used in the previous analysis and explained in Sect. 4.1.

Working with the $E^{dep.}$ analysis variable in the near and far detectors, exactly the same procedure can be followed to produce a PRISM prediction of the FD event rate as set out in Chap. 4. Examples of the PRISM predictions for each signal channel produced using $E^{dep.}$ are shown in Fig. 7.1. The full oscillation analysis presented in Chap. 6 can be performed using deposited energy.

7.2 Missing Proton Energy Mock Data

A mock data set has been produced by members of the DUNE Collaboration in order to test the DUNE oscillation measurement in a scenario in which the MC simulation does not accurately predict the data. Unobserved particles, such as neutrons, may carry away significant amounts of energy in a neutrino interaction. The mock data set therefore scales down the deposited energy of protons by 20%, assuming that this energy is carried away by the undetected particles. Hence, the deposited energy of each event in the ND and FD mock data is scaled such that

$$E^{dep.} \rightarrow E^{dep.\prime} = E^{dep.} - 0.2 \times E^{dep.}_{p}. \tag{7.2}$$

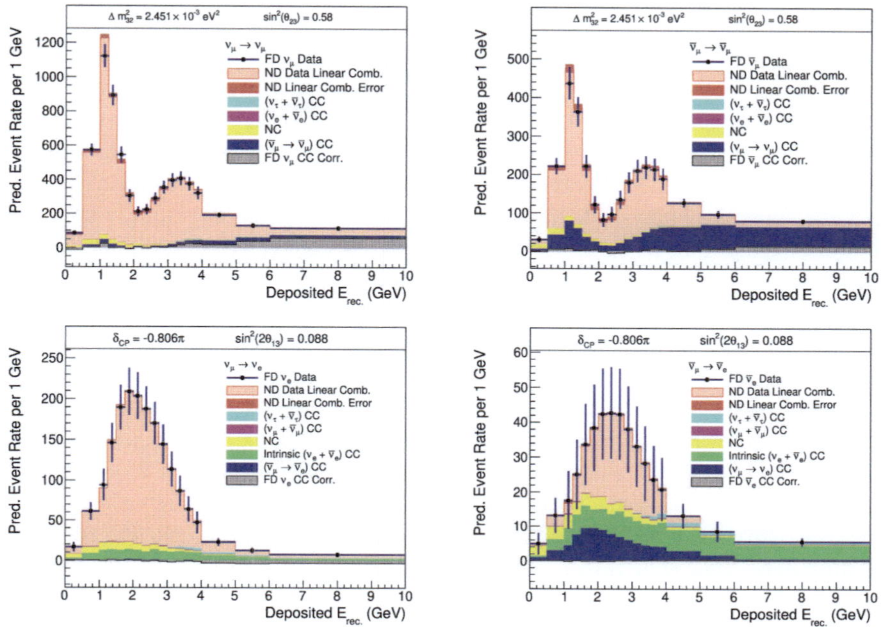

Fig. 7.1 PRISM predictions for the four DUNE signal channels made in the variable $E^{rec.}_{dep.}$. The total stacked histogram is the PRISM prediction. The methodology used to produce the predictions is detailed in Chap. 4

In an On-Axis DUNE oscillation analysis, the MC simulation will attempt to predict the mock data. A fit at the ND is used to vary the parameters of the neutrino interaction model until a good prediction of the mock data is achieved. This fit will not necessarily choose the correct way to alter the neutrino interaction model, which would be to reduce the proton deposited energy by 20%. Worse, it is plausible that no tuning of the existing model will provide a satisfactory fit of the ND MC to the on-axis ND data, leading to the inclusion of additional empirical parameters.

An analogous scenario is established in which the on-axis ND mock data, which has had 20% of the proton energy removed according to Eq. 7.2, is reweighted to appear identical to the on-axis ND nominal MC in deposited energy. The alternative option would be to reweight the on-axis ND nominal MC so that it matched the on-axis ND mock data. The former option is chosen, although both cases are equivalent in simulating a scenario in which the neutrino interaction model is altered to achieve a good-fit between the on-axis ND MC and ND data in deposited energy.

Section 4.3.3 of Ref. [1] demonstrates that the on-axis ND mock data can reproduce the nominal MC prediction in deposited energy ($E^{dep.}$) through a multivariate reweighting method [2]. Despite the reweighted mock data recovering the nominal MC prediction in $E^{dep.}$, Ref. [1] shows that the relationship between true and reconstructed neutrino energy is biased. The reweighting scheme is then extrapolated to the ND mock data at all off-axis positions and the FD mock data. Further details of this procedure can be found in Sect. 4.3.4 of Ref. [1]. Once extrapolated to the oscillated neutrino flux at the DUNE FD, the biased relationship between true and reconstructed neutrino energy results in poor agreement between the FD mock data and the MC prediction. Different neutrino fluxes are also measured at the ND off-axis positions, where discrepancies between the ND nominal MC and mock data due to a poor choice of alterations to the neutrino interaction model are likely to become apparent. Hence, PRISM is a powerful tool in disentangling the convolution of neutrino flux and cross section that may obscure deficiencies in the chosen parameterisation of the neutrino interaction model.

In summary, the proton deposited energy in the ND and FD data is suppressed according to Eq. 7.2. The on-axis ND data is then reweighted and the reweighting scheme is extrapolated to the FD mock data and ND off-axis samples. This results in a mock data set at the ND and FD, where the ND on-axis mock data is in good agreement with the nominal MC, but the FD and off-axis ND mock data show disagreements with the nominal MC. This is illustrated in Figs. 7.2 and 7.3 for FHC and RHC mode respectively. The impact of the reduction in deposited proton energy is smaller in RHC mode since antineutrino interactions tend to produce a neutron in the final state rather than a proton.

7.2.1 Mock Data Impact on Oscillation Measurement

A clear shift can be seen between the nominal FD MC and the FD mock data in Fig. 7.2. However, since the on-axis ND data is in good agreement with the nominal MC, the On-Axis analysis will constrain the flux and cross section parameters to be close to their initial values in the fit of the nominal MC to the on-axis ND data. Consequently, the oscillation parameters are altered to the wrong values in order to achieve a FD MC prediction in good agreement with the FD mock data. This is illustrated in Fig. 7.4, taken from a study in Ref. [1], where the application of the missing proton energy mock data reweighting causes a clear bias in the On-Axis analysis contours. Neither the freedom to vary the systematic uncertainties in the fit nor increasing the exposure are able to reduce the bias in the On-Axis contours. The next section shows how a PRISM analysis avoids such a bias in the oscillation contours when using the same mock data samples in the ND and FD.

7.3 PRISM Oscillation Measurement with Mock Data

The near and far detector "data" is shifted and reweighted according to the procedure set out in Sect. 7.2. A full PRISM oscillation parameter measurement is then performed following exactly the same procedure as outlined in Chap. 6. Initially, systematic uncertainties are not accounted for in the oscillation fit. The PRISM analysis linearly combines the ND mock data at many off-axis positions to build a prediction

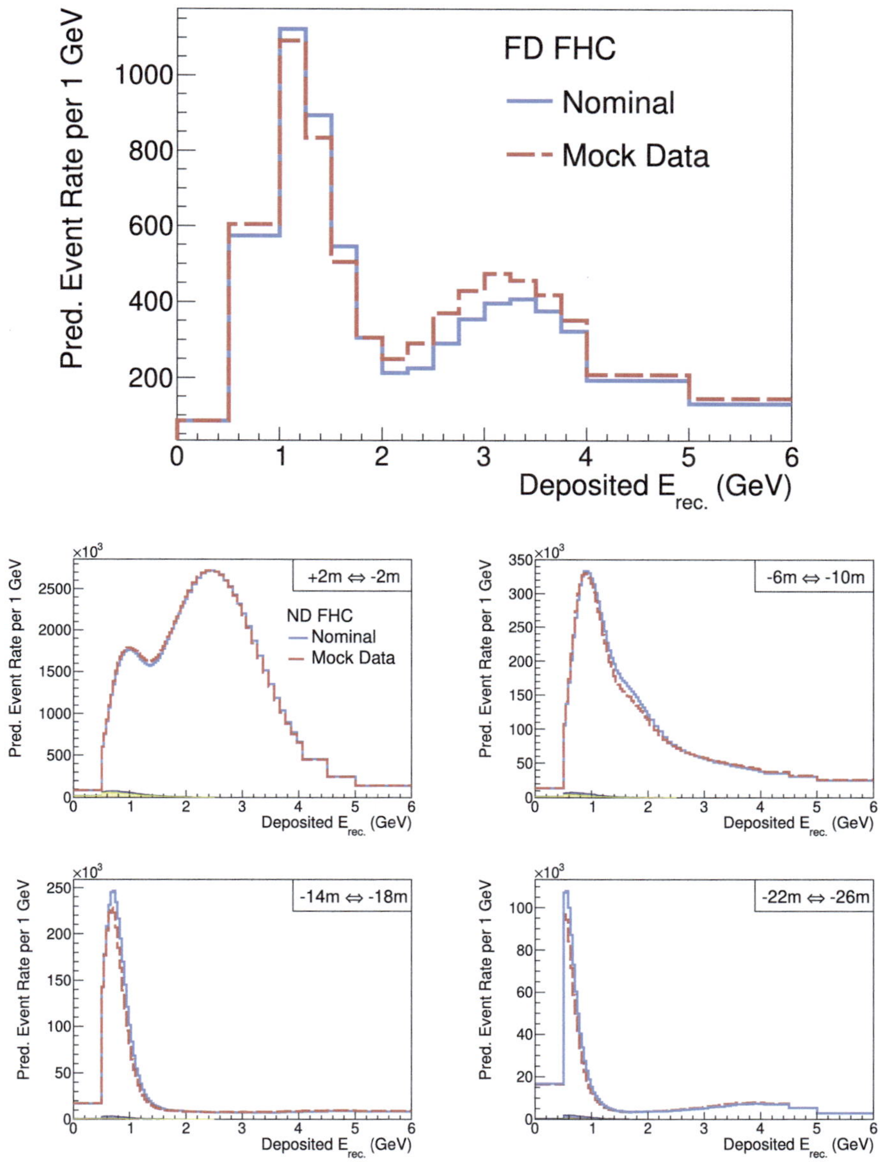

Fig. 7.2 Event rate predictions for ν_μ according to the nominal MC and mock data for the FD (top) and at different off-axis positions at the ND (bottom) in FHC mode. The ND event rates are integrated over the off-axis range indicated in the corresponding legend. The small stacked histograms are the ND NC and wrong-sign backgrounds. There is a minimum at approximately 1 GeV in the ND event rates, which is due to the exclusion of events where the muon is neither contained nor propagates to TMS/ND-GAr

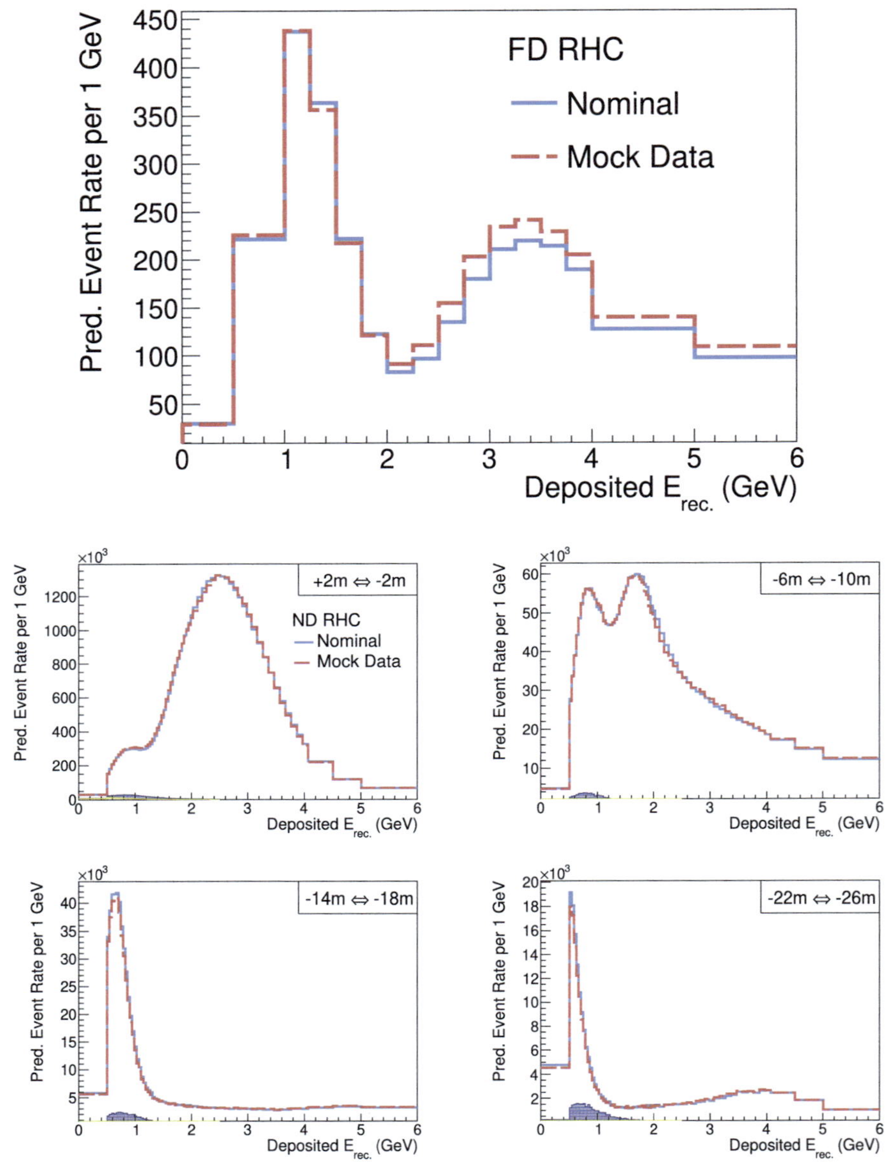

Fig. 7.3 Event rate predictions for $\bar{\nu}_\mu$ according to the nominal MC and mock data for the FD (top) and at different off-axis positions at the ND (bottom) in RHC mode. The selected ND event rates are integrated over the off-axis range indicated in the corresponding legend. The small stacked histograms are the ND NC and wrong-sign backgrounds. There is a minimum at approximately 1 GeV in the ND event rates, which is due to the exclusion of events where the muon is neither contained nor propagates to TMS/ND-GAr

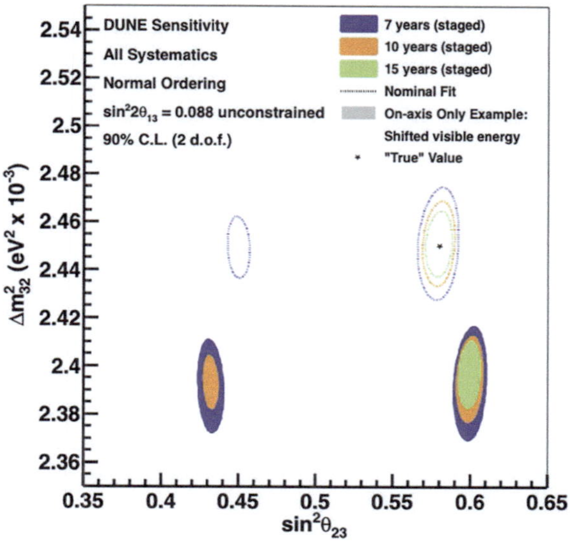

Fig. 7.4 Contours for fits of the nominal FD MC to the nominal FD MC (dashed contours) and to the mock data (solid contours). A joint fit at the FD and on-axis ND is performed that accounts for all sources of systematic uncertainty. The fit at the on-axis ND constrains the systematic uncertainty parameters. The true oscillation point is indicated by the star and 90% contours for three exposures are shown. Figure taken from Ref. [1]

of the FD mock data. The MC components, such as the background predictions, are derived from the nominal MC. The purpose of this study is to examine whether the PRISM analysis can perform an accurate measurement of the oscillation parameters despite the nominal MC providing an inaccurate mapping between the true and reconstructed quantities in a neutrino interaction. It was seen in the previous section that this is not always possible in the On-Axis DUNE analysis.

Following the methodology detailed in Chap. 6, a joint fit of PRISM predictions to the FD mock data for the four signal channels is performed. The PRISM contours resulting from this fit in a Δm^2_{32} versus $\sin^2\theta_{23}$ parameter space are shown in Fig. 7.5. In addition, Fig. 7.5 provides contours for a fit of the FD MC to the FD mock data. This is equivalent to the On-Axis analysis measurement, but without the ND fit to constrain systematic uncertainty parameters. Figure 7.5 demonstrates that, whilst the PRISM oscillation measurement for Δm^2_{32} and $\sin^2\theta_{23}$ is not as biased as the On-Axis analysis measurement, a significant bias in the PRISM contours does remain.

The cause of the bias in Fig. 7.5 can be clearly seen by comparing the PRISM predictions to the mock FD data. Figure 7.6 shows the total PRISM prediction produced by linearly combining off-axis ND mock data and the FD mock data assuming the same NuFIT 4.0 oscillation parameters. In order to make an unbiased measurement of the oscillation parameters, the PRISM prediction should agree well with the FD mock data when they are produced with the same oscillation parameters. However, Fig. 7.6 shows poor agreement between the PRISM prediction and FD mock data. The fit of the PRISM prediction to the FD mock data therefore changes the oscillation parameters to incorrect values in order to force agreement between the prediction and mock data, resulting in the biased contours seen in Fig. 7.5.

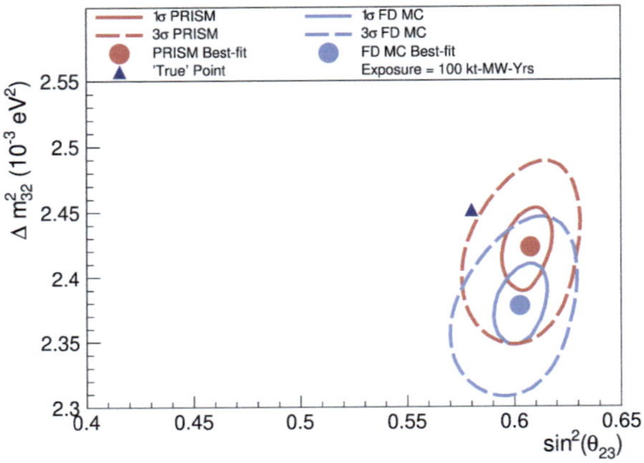

Fig. 7.5 Four signal channel fits of the PRISM (red) and FD MC (blue) predictions to the FD mock data in a Δm^2_{32} versus $\sin^2\theta_{23}$ parameter space. 1σ and 3σ contours are plotted for a single exposure of 100 kt-MW-Yrs

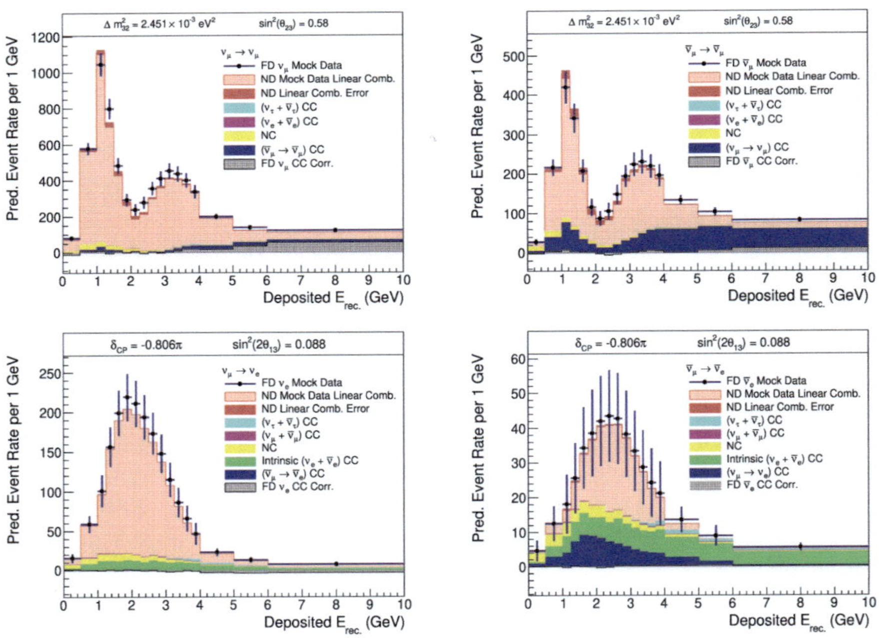

Fig. 7.6 FD mock data and PRISM predictions produced by linearly combining ND mock data for the channels $\nu_\mu \to \nu_\mu$ (top left), $\bar{\nu}_\mu \to \bar{\nu}_\mu$ (top right), $\nu_\mu \to \nu_e$ (bottom left) and $\bar{\nu}_\mu \to \bar{\nu}_e$ (bottom right). There is poor agreement between the PRISM predictions and FD mock data in both channels

7.3.1 Data-Driven Geometric Efficiency Correction

Upon examining the impact of the cross section uncertainties on the PRISM analysis, which was studied in detail in Chaps. 5 and 6, it should not be surprising that the PRISM contours in Fig. 7.5 show a bias relative to the true oscillation point. It has been repeatedly shown that MC-based components of the PRISM prediction introduce neutrino interaction model dependencies into the PRISM analysis. The MC-based components are derived from the nominal MC simulation, which no longer accurately describes the near and far detector data due to the suppression of the deposited proton energy and reweighting procedure. In particular, Sects. 5.2.2 and 6.6.2 demonstrated that the ND efficiency correction is the primary MC component through which cross section systematic uncertainties enter the PRISM analysis. It was shown in Fig. 6.12 that the inclusion of a data-driven ND efficiency correction has the potential to drastically reduce the impact of cross section uncertainties. The data-driven efficiency correction would be the geometric efficiency correction described in detail in Sect. 4.7.1.

A similar study is performed in order to determine the impact of the ND efficiency correction on the mock data oscillation fits. Since a data-driven ND efficiency correction would be derived directly from the ND mock data, the mock data proton energy shift and reweighting procedure is applied to the MC used to calculate the ND efficiency. The ND efficiency correction is therefore based on the correct neutrino interaction physics present in the data, rather than the incorrect model in the nominal MC. A four signal channel joint fit of the PRISM prediction to the FD mock data is performed again in a Δm^2_{32} versus $\sin^2\theta_{23}$ parameter space. Contours for this fit are shown in Fig. 7.7. The bias in the PRISM contours has been substantially reduced relative to the PRISM contours shown in Fig. 7.5 and this is entirely due to the expected impact of ND geometric efficiency correction.

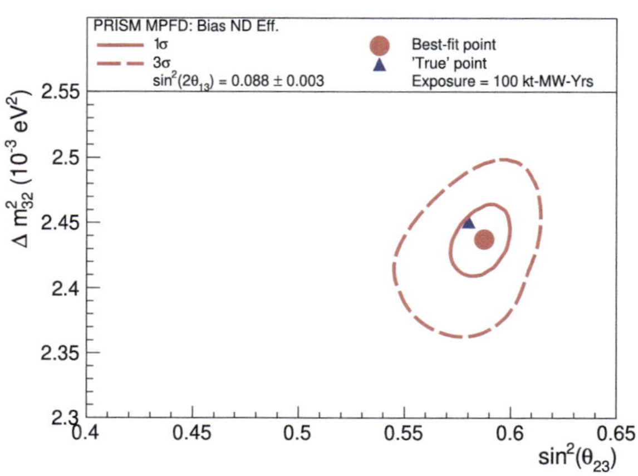

Fig. 7.7 Four signal channel fit of the PRISM prediction to the FD mock data in a Δm^2_{32} versus $\sin^2\theta_{23}$ parameter space. The mock data proton energy shift and reweighting scheme has been applied to the ND efficiency calculation for the PRISM prediction. 1σ and 3σ contours are plotted for a single exposure of 100 kt-MW-Yrs

7.3.2 Wrong-Sign Background

Whilst the bias in the PRISM contours is much reduced in Fig. 7.7, where the true oscillation point lies within the 1σ contours at an exposure of 100 kt-MW-Yrs, some bias still remains. The source of the remaining bias in the oscillation contours can be deduced by performing separate joint fits for the FHC and RHC signal channels. Figure 7.8 shows contours in a Δm^2_{32} versus $\sin^2\theta_{23}$ parameter space for separate FHC ν_μ and ν_e and RHC $\bar{\nu}_\mu$ and $\bar{\nu}_e$ joint fits, where the mock data proton energy shift and reweighting procedure has been applied to the ND efficiency calculations, as described in Sect. 7.3.1. A clear difference can be seen in the degree of bias between the FHC and RHC fit. The FHC joint fit has almost no bias in the oscillation contours, however the RHC contours have a significant amount of bias remaining. There is only one MC component of the PRISM prediction that differs significantly between the FHC and RHC signal channels: the FD wrong-sign background. Figure 7.1 shows that the wrong-sign background forms a large component of the RHC $\bar{\nu}_\mu$ prediction. In the current PRISM analysis this component is entirely derived from the nominal MC.

It is concluded from Fig. 7.8 that the presence of a FD wrong-sign background component derived from the nominal MC is the primary cause of the remaining bias in the PRISM oscillation measurement in a four signal channel fit. This can be further demonstrated by applying the mock data proton energy shift and reweighting procedure to the FD wrong-sign background in addition to the ND efficiency calculation. Doing so simulates a scenario in which both the FD wrong-sign background and the ND efficiency correction are derived from the data rather than the nominal MC. These are improvements that are planned for the PRISM analysis and a data-driven method for predicting FD wrong-sign background is described in Sect. 4.7.2. Separate fits for the FHC and RHC signal channels are then repeated and the resulting contours are shown in Fig. 7.9. The FHC contours have changed little between Figs. 7.8 and 7.9, however the bias in the RHC contours has been almost completely removed.

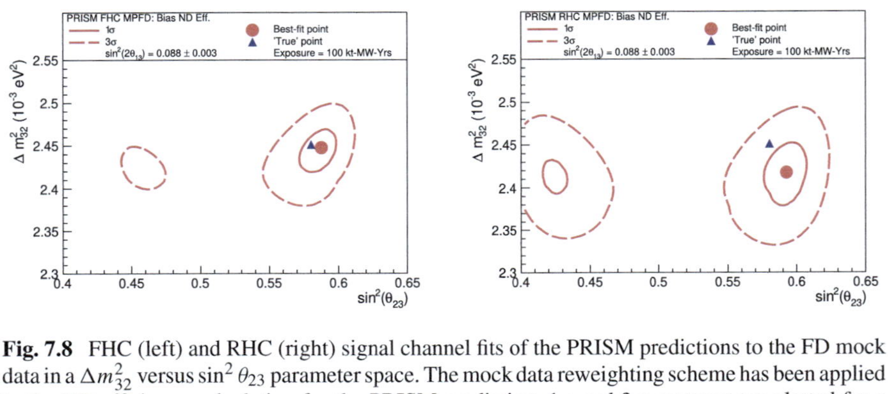

Fig. 7.8 FHC (left) and RHC (right) signal channel fits of the PRISM predictions to the FD mock data in a Δm^2_{32} versus $\sin^2\theta_{23}$ parameter space. The mock data reweighting scheme has been applied to the ND efficiency calculation for the PRISM prediction. 1σ and 3σ contours are plotted for a single exposure of 100 kt-MW-Yrs

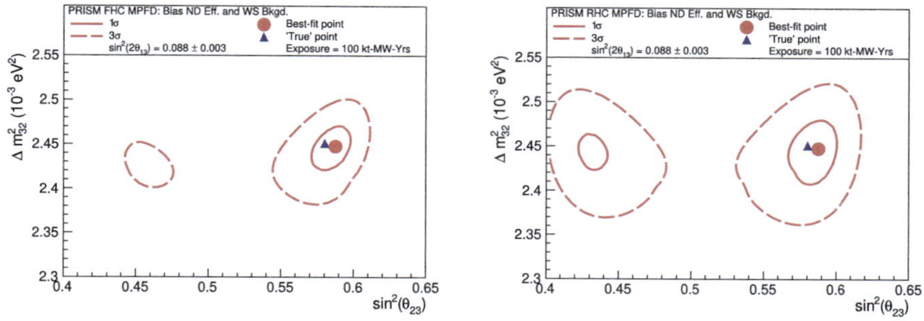

Fig. 7.9 FHC (left) and RHC (right) signal channel fits of the PRISM predictions to the FD mock data in a Δm^2_{32} versus $\sin^2\theta_{23}$ parameter space. The mock data reweighting scheme has been applied to the ND efficiency calculation and the FD wrong-sign background for the PRISM prediction. 1σ and 3σ contours are plotted for a single exposure of 100 kt-MW-Yrs

With the mock data proton energy shift and reweighting applied to the FD wrong-sign background and ND efficiency calculation, a four signal channel PRISM joint oscillation fit to the FD mock data is performed in Δm^2_{32} versus $\sin^2\theta_{23}$ parameter space. Figure 7.10 shows the PRISM contours for such a fit, where the true oscillation point clearly lies in the centre of the 1σ contours. This demonstrates that, despite the nominal MC providing an inaccurate mapping between the true neutrino energy and reconstructed energy of a neutrino interaction, a PRISM linear combination analysis can produce an accurate measurement of the oscillation parameters with minimal bias. However, this is only possible provided the calculation of the ND selection efficiency and the prediction of the FD wrong-sign background are derived from the data, rather than the MC, and therefore contain the correct neutrino interaction physics. Both of these features are achievable and are discussed in detail in Sect. 4.7.

It is useful to again look at the PRISM predictions and FD mock data to demonstrate why a bias in the oscillation contours is no longer expected. Figure 7.11 compares FD mock data and the corresponding PRISM predictions produced by linearly combining off-axis ND mock data at the NuFIT 4.0 oscillation parameters. However, unlike in Fig. 7.11, the mock data reweighting scheme has been applied to the ND efficiency calculation and the FD wrong-sign background. This time, using the same set of oscillation parameters, good agreement can be seen between the PRISM prediction and FD mock data. Therefore, during a fit of the PRISM prediction to the FD mock data, the best fit is expected to be found very close to the true oscillation point, resulting in the unbiased contours seen in Fig. 7.10.

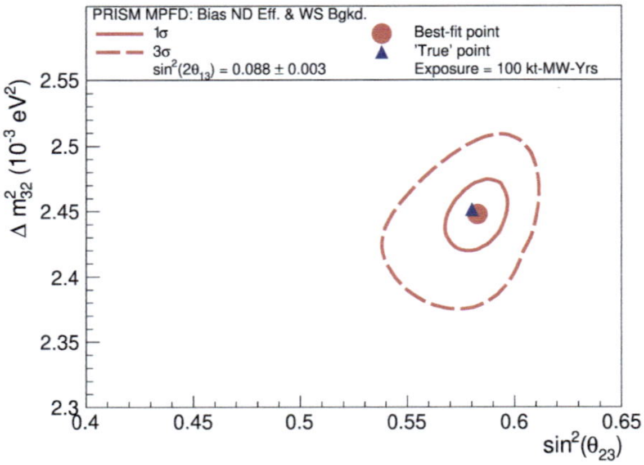

Fig. 7.10 Four signal channel fit of the PRISM prediction to the FD mock data in a Δm^2_{32} versus $\sin^2\theta_{23}$ parameter space. The mock data proton energy shift and reweighting scheme has been applied to the ND efficiency calculation and the FD wrong-sign background for the PRISM prediction. 1σ and 3σ contours are plotted for a single exposure of 100 kt-MW-Yrs

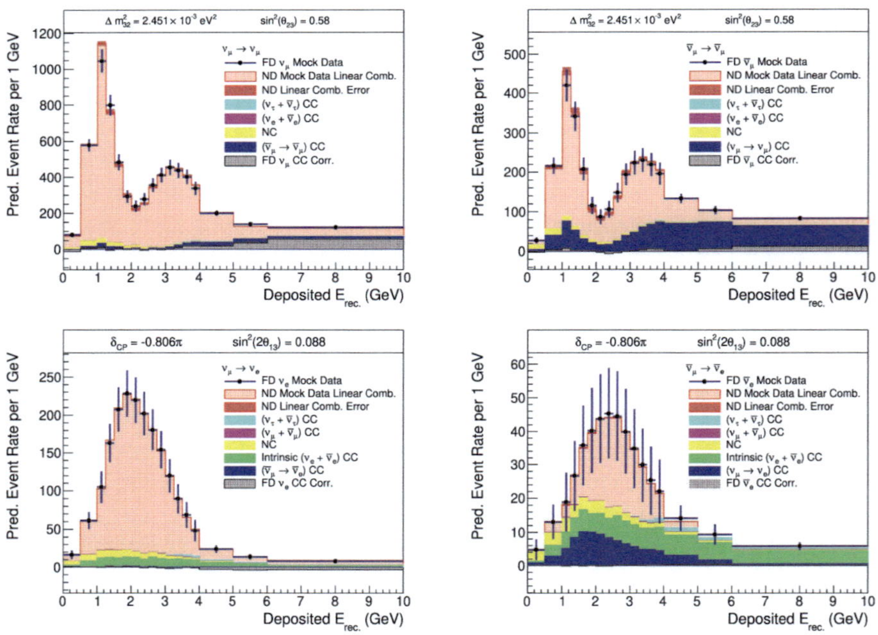

Fig. 7.11 FD mock data and PRISM predictions produced by linearly combining ND mock data for the channels $\nu_\mu \rightarrow \nu_\mu$ (top left), $\bar{\nu}_\mu \rightarrow \bar{\nu}_\mu$ (top right), $\nu_\mu \rightarrow \nu_e$ (bottom left) and $\bar{\nu}_\mu \rightarrow \bar{\nu}_e$ (bottom right). The mock data proton energy shift and reweighting procedure has now been applied to the ND efficiency calculation and wrong-sign background components of the PRISM predictions. Good agreement has been recovered between the PRISM predictions and FD mock data in all channels

7.4 Mock Data Fits Accounting for Systematic Uncertainty

The PRISM oscillation analysis using mock data at the ND and FD can be developed further with the inclusion of systematic uncertainties in the oscillation fits. Following the procedure of the previous sections, the mock data proton energy shift and reweighting scheme is applied to the ND efficiency calculation and the FD wrong-sign background as well as the near and far detector data. The PRISM prediction, built from linearly combined ND mock data, is then fitted to the FD mock data, accounting for all the flux, cross section and detector systematic uncertainties in the joint fit. Because this scenario simulates the impact of including a data-driven ND efficiency correction and FD wrong-sign background, these components are not allowed to be varied by systematic uncertainties in the fit. This is an over-simplification, since the linear combination prediction of the wrong-sign background will be affected by the flux and detector uncertainties. However, since the sensitivity is driven by the FHC predictions where the wrong-sign background is small, this is not expected to be a dominant effect. The same FD staging plan and ND run-plan established in Sect. 6.1 is assumed.

The flux, cross section and detector systematic uncertainties included in the fit are identical to those defined in Chap. 5 and used in the Chap. 6 oscillation fits. The only difference in the systematic uncertainty implementation from previous chapters is that, due to the use of the true lepton energy in the definition of $E^{dep.}$ in Eq. 7.1, there is no uncertainty on the lepton energy resolution.

Figure 7.12 shows single-parameter χ^2 curves for Δm^2_{32}, $\sin^2\theta_{23}$ and δ_{CP} at three different exposures and accounting for all the flux, cross section and detector systematic uncertainties. It can be seen that, despite the mock data at the FD, the PRISM analysis is able to measure the correct oscillation parameters without bias by producing predictions of the FD mock data from linearly combined ND mock data.

Contours are then displayed at a single exposure in two-dimensional oscillation parameter spaces in Fig. 7.13. As in the previous figure, the impact of all the flux, cross section and detector uncertainties are accounted for in the oscillation fits and the 1σ and 3σ contours are plotted. The θ_{13} reactor constraint is removed for the $\sin^2 2\theta_{13}$ versus δ_{CP} sensitivity. Figure 7.13 again demonstrates that no bias exists even when including systematic uncertainties in a scenario where the neutrino interaction model provides a poor description of the data.

Figures 7.12 and 7.13 demonstrate that the PRISM oscillation analysis can achieve its primary purpose, which is to measure the oscillation parameters with minimal dependence on the neutrino interaction model. This is significant because, as shown in Sect. 7.2.1, mismodelling of neutrino interactions can lead to biases in the standard "On-Axis" DUNE oscillation parameter measurement. The PRISM oscillation analysis can greatly reduce the risk of such a scenario occurring, providing oscillation sensitivities that are robust against neutrino interaction modelling errors. Data-driven methods for calculating the ND geometric efficiency and predicting the FD wrong-sign background are necessary in order to sufficiently suppress the neutrino interaction model dependence of the PRISM method. However, at the time of writing

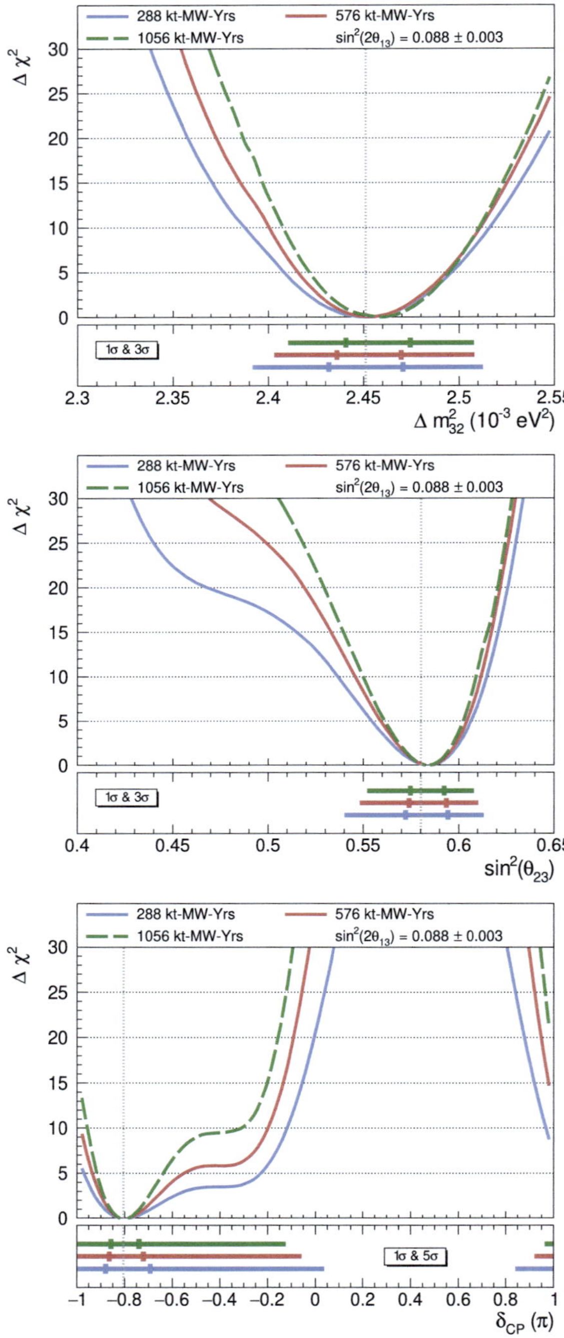

Fig. 7.12 Four signal channel fits of the PRISM predictions to the FD mock data for Δm^2_{32}, $\sin^2\theta_{23}$ and δ_{CP}. All flux, cross section and detector systematic uncertainties are accounted for in the fits. The mock data reweighting scheme has been applied to the ND efficiency calculation and the FD wrong-sign background for the PRISM predictions

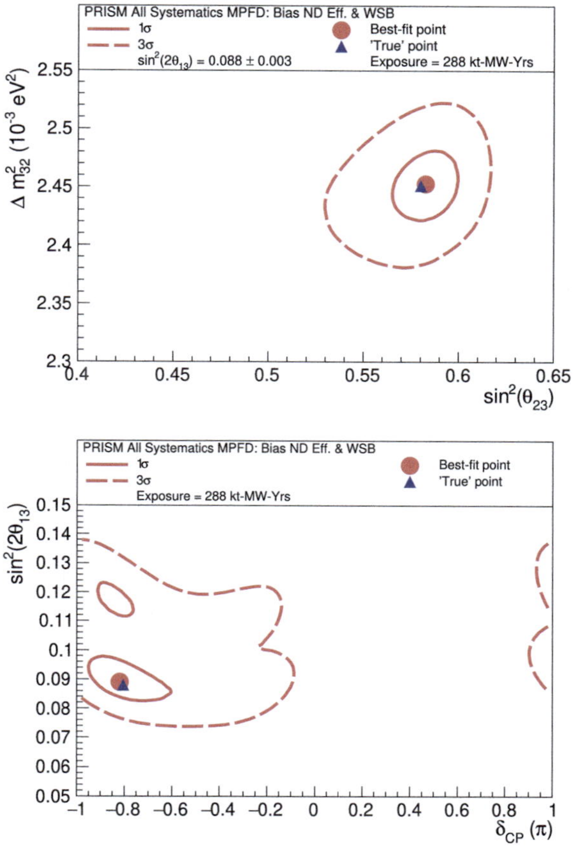

Fig. 7.13 PRISM oscillation contours at a fixed exposure in a two-dimensional oscillation parameter spaces. The reactor neutrino θ_{13} constraint is removed for the $\sin^2 2\theta_{13}$ versus δ_{CP} sensitivity. All flux, cross section and detector systematic uncertainties are accounted for in the fits. The mock data reweighting scheme has been applied to the ND efficiency calculation and the FD wrong-sign background for the PRISM predictions

plans for including such features are at an advanced stage and details of how these additional data-driven methods will be implemented are provided in Sect. 4.7. With the addition of these features, the PRISM analysis has the potential to deliver the most model independent measurement to-date of the neutrino oscillation parameters in a long baseline neutrino oscillation experiment.

References

1. DUNE Collaboration, A.A. Abud, et al., Deep underground neutrino experiment (DUNE) near detector conceptual design report. Instruments **5**(4), 31 (2021). arXiv: 2103.13910 [physics.ins-det]
2. A. Rogozhnikov, Reweighting with boosted decision trees. J. Phys. Conf. Ser. **762**(1) (2016). arXiv:1608.05806 [physics.data-an]. (L. Salinas, C. Torres (eds.), pp. 012–036)

Chapter 8
Conclusions and Outlook

The purpose of this thesis was to detail the methodology, discuss the potential challenges and demonstrate the expected advantages of a PRISM neutrino oscillation analysis for DUNE. The DUNE-PRISM system is the ability to move the ND to different off-axis positions relative to the neutrino beam axis. However these off-axis measurements are ultimately used, it is anticipated that PRISM will be a key component of the DUNE physics program. As such, this thesis is the first to explain in detail how this feature of the DUNE ND can be used to perform a neutrino oscillation analysis for DUNE. The work presented in this thesis represents a major advance on the PRISM disappearance predictions presented in the ND Conceptual Design Report [1] and has brought the PRISM analysis to the same level of sophistication as the On-Axis oscillation analysis published in the FD Technical Design Report [2].

In order to perform a PRISM oscillation measurement, a data-driven prediction of the FD event rate is first needed. The method to produce a PRISM prediction for each of the four signal channels was described in Chap. 4. This is an involved process that requires several steps to achieve the desired result. The method for correcting for differences between the near and far detectors (Sect. 4.3) was developed for this thesis and allowed a realistic PRISM oscillation measurement to be performed using reconstructed variables and an event selection at both detector sites. The PRISM method described in Chap. 4 requires the MC simulation at several stages, introducing model-dependence into the PRISM prediction. A number of these MC components are regarded as temporary features that will eventually be replaced by data-driven methods. Section 4.7 explained the various methodology improvements that are planned for the PRISM analysis. The most important of these improvements is the development of a data-driven geometric efficiency correction for the ND.

The DUNE sensitivity is expected to be limited by the impact of the systematic uncertainties, not by the sample size at the FD. This was demonstrated in Fig. 6.25, which illustrated how further improvements in the oscillation parameter sensitivity become small at large FD sample sizes. A discussion on the flux, cross section

C. M. Hasnip, *DUNE-PRISM—A New Method to Measure Neutrino Oscillations*, Springer Theses, https://doi.org/10.1007/978-3-032-07251-1_8

and detector systematic uncertainties and their potential impact began in Chap. 5. Since the PRISM method relies on the flux model to linearly combine the ND data correctly, particular attention was paid to the impact of flux systematic uncertainties on the PRISM prediction. Whilst a degree of cancellation of the flux systematic uncertainties between the near and far detectors was demonstrated, a number of potentially problematic focusing, horn alignment and beam alignment uncertainty parameters have been identified. A series of error band studies then indicated that the flux uncertainties largely entered the PRISM analysis through the linear combination coefficient calculation and the cross section uncertainties can have a large influence on the PRISM prediction through the MC-based ND efficiency correction. Finally, it was shown in Sect. 5.3 that the detector uncertainties have a significant effect on the PRISM prediction and largely enter the analysis through the smearing matrices that correct for resolution differences between the ND and FD.

A realistic PRISM oscillation analysis including the four signal channels and all flux, cross section and detector systematic uncertainties was presented in Chapter 6. This chapter focused on "Asimov-like" studies of the PRISM sensitivity to Δm^2_{32}, $\sin^2\theta_{23}$, δ_{CP} and $\sin^2 2\theta_{13}$. In addition, the discussion on the relative impact of the different systematic uncertainties from Chap. 5 was further developed. Section 6.6.1 confirmed that the flux uncertainties primarily enter the analysis through the calculation of the linear combination coefficients and identified the four flux systematic parameters that are particularly problematic for PRISM: "Horn Current", "Decay Pipe Radius", "Horn 1 X-Shift" and "Beam X-Offset". The importance of the ND efficiency correction as the primary way cross section systematic uncertainties enter the analysis was again highlighted in Sect. 6.6.2. Section 6.7 and Fig. 6.16 demonstrated the improvement in the PRISM sensitivity that can be achieved if a data-driven correction for the ND geometric efficiency were to be included in the analysis. This chapter finished with direct comparisons between the PRISM and On-Axis analysis sensitivities, concluding that the PRISM analysis is currently less sensitive to Δm^2_{32} and $\sin^2\theta_{23}$ than the On-Axis analysis and this is largely due to the significant impact of the flux systematic uncertainties on the PRISM linear combination.

Chapter 6 presented a series of "Asimov-like" studies where the nominal MC neutrino interaction physics was the same as the "data". In a real DUNE measurement this will not be the case. Hence, Chap. 7 explored a scenario where the nominal MC provides a poor description of the data and determined how a PRISM measurement would cope with such circumstances. A mock data sample was introduced that shifted the proton kinetic energy and reweighted each event to create a data-set in the ND and FD that, by construction, agreed well with the on-axis ND nominal MC, but fitted poorly to the FD nominal MC. This simulated a plausible scenario where the neutrino interaction model is adjusted to achieve good agreement between the on-axis ND data and MC, but the mapping between the true and reconstructed neutrino energy is biased. Referencing a previous study [1], Sect. 7.2.1 demonstrated that such a scenario can lead to a bias in the oscillation parameter measurement. Section 7.3 then explained how a PRISM measurement, with the inclusion of a data-driven ND geometric efficiency correction and a data-driven wrong-sign background, can largely remove the bias in the oscillation parameter measurement. This is significant as it demonstrates the PRISM oscillation analysis can achieve its primary purpose, which is to robustly measure the neutrino oscillation parameters with minimal dependence on the neutrino interaction model.

As discussed in Sect. 4.7, there are a number of improvements to be made to the PRISM methodology and further studies are needed into the impact of the various systematic uncertainties. The priorities for improving the PRISM analysis are as follows:

- Implement the data-driven ND geometric efficiency correction in the PRISM methodology, replacing the current MC-based correction. An alternative approach would be to remove off-axis positions from the PRISM linear combination that require large MC-based efficiency corrections. This is the most important way neutrino interaction model dependence can be removed from the PRISM analysis.
- Include data-driven predictions of the FD wrong-sign and intrinsic $\nu_e/\bar{\nu}_e$ backgrounds. In particular, the data-driven wrong-sign prediction has been shown to be vital to reducing the neutrino interaction model dependence of the RHC PRISM predictions. These features have already been implemented in the PRISM software by members of the DUNE-PRISM working group.
- Further study is needed into all the possible sources of uncertainty in the neutrino flux using an updated simulation of the LBNF beam and an expanded list flux parameters. If a small group of problematic flux parameters can be identified, then it may be possible to focus on reducing the uncertainty in specific components of the flux simulation.
- Explore the measurement of the $\sigma(\nu_e)/\sigma(\nu_\mu)$ using the PRISM method. The appearance measurement would then not rely solely on the MC prediction of the $\sigma(\nu_e)/\sigma(\nu_\mu)$ ratio when predicting the electron neutrino event rate.

In summary, this thesis proves that PRISM can be used to measure the full suite of oscillation parameters accessible by DUNE in a manner that is minimally dependent on the neutrino interaction model. In addition, a complete treatment of the available flux, cross section and detector systematic uncertainties has been implemented. PRISM is now a key component of the DUNE physics program and it is hoped that this thesis can be a helpful guide to any neutrino physicist that wishes to further develop this data-driven oscillation analysis.

References

1. DUNE Collaboration, A.A. Abud, et al., Deep underground neutrino experiment (DUNE) near detector conceptual design report. Instruments **5**(4), 31 (2021). arXiv: 2103.13910 [physics.ins-det]
2. DUNE Collaboration, B. Abi, et al., Deep underground neutrino experiment (DUNE), far detector technical design report, vol. II. DUNE Physics (Feb 2020). arXiv: 2002.03005 [hep-ex]

Appendix
Tikhonov Regularisation Optimisation

Tikhonov regularisation is method for solving ill-posed linear inverse problems that suppresses the variances of the solution by applying a regularisation condition. The method was developed independently by Phillips and Tikhonov [1, 2]. Chapter 4, Sect. 4.4 explains how the linear combination coefficients are calculated through Tikhonov regularisation by minimising

$$\mathcal{L} = \mathcal{L}_1 + \mathcal{L}_2, \tag{A.1}$$

where,

$$\mathcal{L}_1 = \|\mathbf{N}c - F\|^2 \tag{A.2}$$

and

$$\mathcal{L}_2 = \|\Gamma c\|^2 . \tag{A.3}$$

N is the matrix of ND fluxes, c is the vector of linear combination coefficients, F is the vector of the FD oscillated flux and Γ is the regularisation matrix. The quantities $\mathcal{L}_1$ and $\mathcal{L}_2$ are often referred to as the *residual norm* and *solution norm* respectively. The form of the Γ matrix was given in Eq. 4.18, which depends on the regularisation parameter τ_{LC}. If τ_{LC} is too small, there will be large fluctuations and negative correlations between adjacent elements of the solution. This would result in a small residual norm, but a large solution norm. Make τ_{LC} too large and the solution will be biased, since the residual norm will become very large. For the form of the regularisation chosen in Eq. 4.18, in the limit where $\tau_{LC} \to \infty$, the linear combination coefficients will tend towards a single constant value. This is because solutions where there are differences between adjacent elements are heavily penalised for large τ_{LC} values.

Consequently, it is necessary to optimise the choice of τ_{LC} for each oscillation signal channel. This done using the "L-Curve" method. The regularisation parameter is varied and the calculated square-root of $\mathcal{L}_1$ and $\mathcal{L}_2$ are plotted against one another.

C. M. Hasnip, *DUNE-PRISM—A New Method to Measure Neutrino Oscillations*, Springer Theses, https://doi.org/10.1007/978-3-032-07251-1

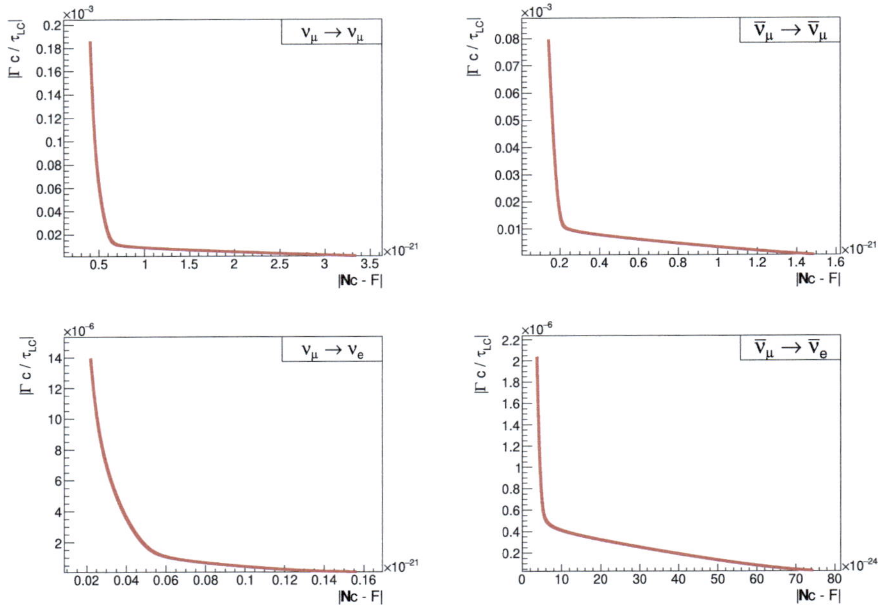

Fig. A.1 A series of "L-Curves" for each of the oscillation channels. The "kink" in the curve corresponds to the optimum regularisation parameter

The $\mathcal{L}_2$ norm is normalised by the regularisation parameter in order to remove any explicit dependence on τ_{LC}. These plots are shown in Fig. A.1 for each of the oscillation parameters. In each figure one can see a characteristic "kink" in the graph where the curvature is maximal. The point at which the curvature is maximal corresponds to the τ_{LC} that simultaneously minimises the residual and solution norms ($\mathcal{L}_1$ and $\mathcal{L}_2$). To move τ_{LC} away from this point would result in either $\mathcal{L}_1$ or $\mathcal{L}_2$ increasing faster than the other decreases. Hence, the point of maximum curvature in the "L-Curve" corresponds to the τ_{LC} that provides the best balance between minimising the bias and suppressing large fluctuations in the solution [3].

The curvature of the "L-Curve" is then calculated and plotted as a function of the regularisation parameter τ_{LC} [3, 4]. The point of maximum curvature is taken to be the optimum regularisation parameter to be used in the PRISM oscillation analysis. Figure A.2 plots the curvature as a function of τ_{LC} for each oscillation parameter. A clear maximum in the curvature can be seen in each case. Finally, the optimum regularisation parameters for each oscillation channel are summarised in Table A.1.

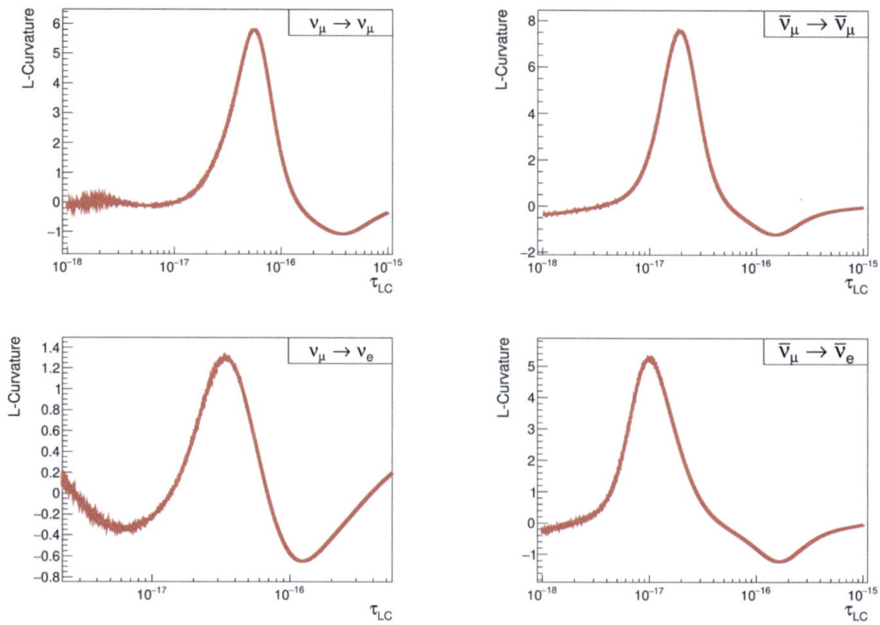

Fig. A.2 Curvature plotted as a function of the regularisation parameter for each oscillation channel. A clear maximum in the curvature can be seen for each oscillation channel, indicating the optimum value for τ_{LC}

Table A.1 Optimum τ_{LC} parameters for each of the signal channels. These are the parameters used for the PRISM analysis presented in this thesis

Channel	Optimum τ_{LC}
$\nu_\mu \to \nu_\mu$	5.58×10^{-17}
$\bar{\nu}_\mu \to \bar{\nu}_\mu$	1.93×10^{-17}
$\nu_\mu \to \nu_e$	3.37×10^{-17}
$\bar{\nu}_\mu \to \bar{\nu}_e$	9.93×10^{-18}

References

1. D.L. Phillips, A technique for the numerical solution of certain integral equations of the first kind. J. Assoc. Comput. Mach. **9**(1), 84–97 (1962)
2. A.N. Tikhonov, V.Y. Arsenin, *Solutions of Ill-Posed Problems* (John Wiley & Sons, Washington, D.C., V. H. Winston & Sons, New York, 1977), pp. xiii258. (Translated from the Russian, Preface by translation editor Fritz John, Scripta Series in Mathematics)
3. P.C. Hansen, The L-curve and its use in the numerical treatment of inverse problems, in *Computational Inverse Problems in Electrocardiology*, vol. 5. Advances in Computational Bioengineering (WIT Press, Southampton, 2001), pp. 119–142
4. S. Schmitt, TUnfold: an algorithm for correcting migration effects in high energy physics. JINST **7**, T10003 (2012). arXiv:1205.6201 [physics.data-an]

MIX
Papier aus verantwortungsvollen Quellen
Paper from responsible sources
FSC® C105338

If you have any concerns about our products,
you can contact us on
ProductSafety@springernature.com

In case Publisher is established outside the EU,
the EU authorized representative is:
Springer Nature Customer Service Center GmbH
Europaplatz 3, 69115 Heidelberg, Germany

Printed by Libri Plureos GmbH
in Hamburg, Germany